'I was riveted by this bo ... of
AI may soon be upon us ... utopian or catastrophic?
The jury is out, but this enlightening, lively and accessible book by a distinguished scientist helps us to assess the odds'
Martin Rees, Astronomer Royal

'Written in an accessible and engaging style, and aimed at the general public, the book offers a political and philosophical map of the promises and perils of the AI revolution'
Yuval Noah Harari, *Guardian*

'This is not just another future-tech book. Artificial intelligence will soon transform work, society, culture and possibly human identity, and Tegmark's excitable yet balanced stance (he is a professor of physics at MIT) makes him an excellent guide'
James McConnachie, *Sunday Times*, Books of the Year

'*Life 3.0* might convince even those who believe that AI is overhyped'
Clive Cookson, *Financial Times*

'In [Tegmark's] magnificent brain, each fact or idea appears to slip neatly into its appointed place like another little silver globe in an orrery the size of the universe. There are spaces for Kant, Cold War history and Dostoyevsky, for the behaviour of subatomic particles and the neuroscience of consciousness . . . Tegmark describes the present, near-future and distant possibilities of AI through a series of highly original thought experiments' Oliver Moody, *The Times*

'Will super-intelligent computers usher in a new era for humanity – or replace us? Very far from a jeremiad against AI. In fact it's much more a celebration of the potential of superintelligence'
Andrew Anthony, *Observer*

'Stands out among the current books about our possible AI futures . . . Tegmark explains brilliantly many concepts in fields from computing to cosmology, writes with intellectual modesty and subtlety, does the reader the important service of defining his terms clearly, and rightly pays homage to the creative minds of science-fiction writers who were, of course, addressing these kinds of questions more than half a century ago. It's often very funny, too' Steven Poole, *Daily Telegraph*

'This is an exhilarating book that will change the way we think about AI, intelligence, and the future of humanity'
Bart Selman, Professor of Computer Science, Cornell University

'Max has written the most insightful and just plain fun exploration of AI's implications that I've ever read. If you haven't been exposed to Max's joyful mind yet, you're in for a huge treat'
Erik Brynjolfsson, co-author of *The Second Machine Age*

'Being an eminent physicist and the leader of the Future of Life Institute has given Max Tegmark a unique vantage point from which to give the reader an inside scoop on the most important issue of our time' Jaan Tallinn, co-founder of Skype

'Max seeks to facilitate a much wider conversation about what kind of future we, as a species, would want to create. Though the topics he covers can be fairly challenging, he presents them in an unintimidating manner that invites the reader to form her own opinions'
Nick Bostrom, author of *Superintelligence*

'Max's new book is a deeply thoughtful guide to the most important conversation of our time'
Ray Kurzweil, author of *The Singularity is Near* and *How to Create a Mind Edit*

ABOUT THE AUTHOR

Max Tegmark is a professor of physics at MIT and president of the Future of Life Institute. He is the author of *Our Mathematical Universe*, and he has featured in dozens of science documentaries. His passion for ideas, adventure and an inspiring future is infectious.

MAX TEGMARK

Life 3.0

Being Human in the Age of Artificial Intelligence

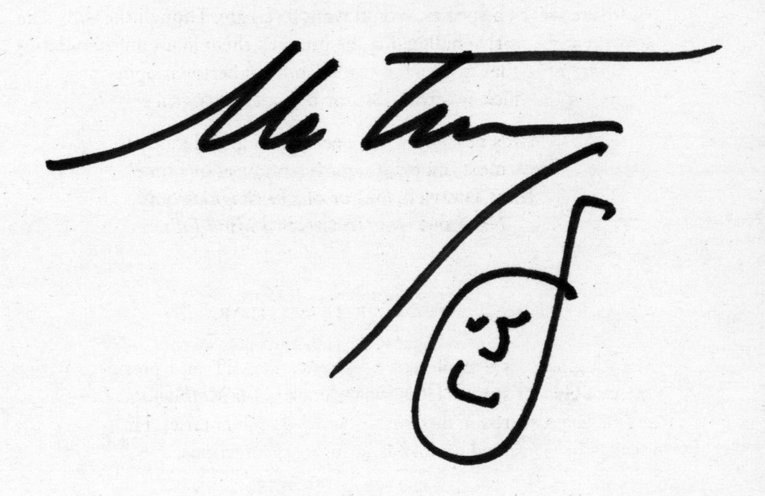

PENGUIN BOOKS

PENGUIN BOOKS

UK | USA | Canada | Ireland | Australia
India | New Zealand | South Africa

Penguin Books is part of the Penguin Random House group of companies
whose addresses can be found at global.penguinrandomhouse.com.

First published in the United States of America by Alfred A. Knopf,
a division of Penguin Random House LLC 2017
First published in Great Britain by Allen Lane 2017
Published in Penguin Books 2018
001

Printed in Great Britain by Clays Ltd, St Ives plc

A CIP catalogue record for this book is available from the British Library

ISBN: 978–0–141–98180–2

www.greenpenguin.co.uk

To the FLI team,
who made everything possible

Contents

Acknowledgments

I'm truly grateful to everyone who has encouraged and helped me write this book, including

my family, friends, teachers, colleagues and collaborators for support and inspiration over the years,
Mom for kindling my curiosity about consciousness and meaning,
Dad for the fighting spirit to make the world a better place,
my sons, Philip and Alexander, for demonstrating the wonders of human-level intelligence emerging,
all the science and technology enthusiasts around the world who've contacted me over the years with questions, comments and encouragement to pursue and publish my ideas,
my agent, John Brockman, for twisting my arm until I agreed to write this book,
Bob Penna, Jesse Thaler and Jeremy England for helpful discussions about quasars, sphalerons and thermodynamics, respectively,
those who gave me feedback on parts of the manuscript, including Mom, my brother Per, Luisa Bahet, Rob Bensinger, Katerina Bergström, Erik Brynjolfsson, Daniela Chita, David Chalmers, Nima Deghani, Henry Lin, Elin Malmsköld, Toby Ord, Jeremy Owen, Lucas Perry, Anthony Romero, Nate Soares and Jaan Tallinn,
the superheroes who commented on drafts of the entire book, namely Meia, Dad, Anthony Aguirre, Paul Almond, Matthew Graves, Phillip Helbig, Richard Mallah, David Marble, Howard Messing,

Luiño Seoane, Marin Soljačić, my editor Dan Frank and, most of all,

Meia, my beloved muse and fellow traveler, for her eternal encouragement, support and inspiration, without which this book wouldn't exist.

LIFE 3.0

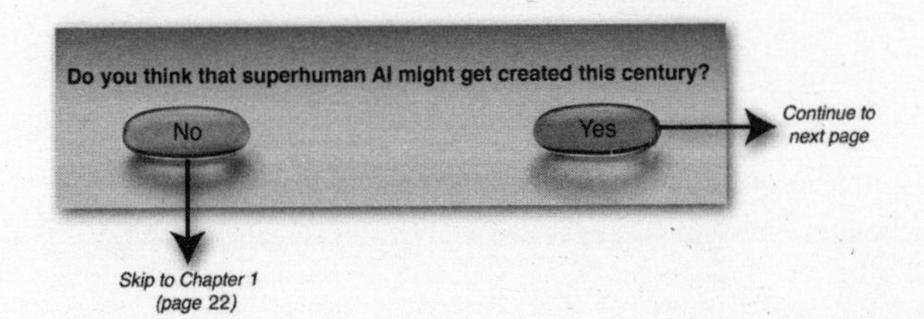
Do you think that superhuman AI might get created this century?
No
Yes
Continue to next page
Skip to Chapter 1 (page 22)

Prelude

The Tale of the Omega Team

The Omega Team was the soul of the company. Whereas the rest of the enterprise brought in the money to keep things going, by various commercial applications of narrow AI, the Omega Team pushed ahead in their quest for what had always been the CEO's dream: building general artificial intelligence. Most other employees viewed "the Omegas," as they affectionately called them, as a bunch of pie-in-the-sky dreamers, perpetually decades away from their goal. They happily indulged them, however, because they liked the prestige that the cutting-edge work of the Omegas gave their company, and they also appreciated the improved algorithms that the Omegas occasionally gave them.

What they didn't realize was that the Omegas had carefully crafted their image to hide a secret: they were extremely close to pulling off the most audacious plan in human history. Their charismatic CEO had handpicked them not only for being brilliant researchers, but also for ambition, idealism and a strong commitment to helping humanity. He reminded them that their plan was extremely dangerous, and that if powerful governments found out, they would do virtually anything—including kidnapping—to shut them down or, preferably, to steal their code. But they were all in, 100%, for much the same reason that many of the world's top physicists joined the Manhattan Project to develop nuclear weapons: they were convinced that if they didn't do it first, someone less idealistic would.

The AI they had built, nicknamed Prometheus, kept getting more capable. Although its cognitive abilities still lagged far behind those of humans in many areas, for example, social skills, the Omegas had pushed hard to make it extraordinary at one particular task: programming AI systems. They'd deliberately chosen this strategy because they had bought the intelligence explosion argument made by the British mathematician Irving Good back in 1965: "Let an ultraintelligent machine be defined as a machine that can far surpass all the intellectual activities of any man however clever. Since the design of machines is one of these intellectual activities, an ultraintelligent machine could design even better machines; there would then unquestionably be an 'intelligence explosion,' and the intelligence of man would be left far behind. Thus the first ultraintelligent machine is the last invention that man need ever make, provided that the machine is docile enough to tell us how to keep it under control."

They figured that if they could get this recursive self-improvement going, the machine would soon get smart enough that it could also teach itself all other human skills that would be useful.

The First Millions

It was nine o'clock on a Friday morning when they decided to launch. Prometheus was humming away in its custom-built computer cluster, which resided in long rows of racks in a vast, access-controlled, air-conditioned room. For security reasons, it was completely disconnected from the internet, but it contained a local copy of much of the web (Wikipedia, the Library of Congress, Twitter, a selection from YouTube, much of Facebook, etc.) to use as its training data to learn from.* They'd picked this start time to work undisturbed: their families and friends thought they were on a weekend corporate retreat. The kitchenette was loaded with microwaveable food and energy drinks, and they were ready to roll.

When they launched, Prometheus was slightly worse than them at

* For simplicity, I've assumed today's economy and technology in this story, even though most researchers guess that human-level general AI is at least decades away. The Omega plan should get even easier to pull off in the future if the digital economy keeps growing and ever more services can be ordered online on a no-questions-asked basis.

programming AI systems, but made up for this by being vastly faster, spending the equivalent of thousands of person-years chugging away at the problem while they chugged a Red Bull. By 10 a.m., it had completed the first redesign of itself, v2.0, which was slightly better but still subhuman. By the time Prometheus 5.0 launched at 2 p.m., however, the Omegas were awestruck: it had blown their performance benchmarks out of the water, and the rate of progress seemed to be accelerating. By nightfall, they decided to deploy Prometheus 10.0 to start phase 2 of their plan: making money.

Their first target was MTurk, the Amazon Mechanical Turk. After its launch in 2005 as a crowdsourcing internet marketplace, it had grown rapidly, with tens of thousands of people around the world anonymously competing around the clock to perform highly structured chores called HITs, "Human Intelligence Tasks." These tasks ranged from transcribing audio recordings to classifying images and writing descriptions of web pages, and all had one thing in common: if you did them well, nobody would know that you were an AI. Prometheus 10.0 was able to do about half of the task categories acceptably well. For each such task category, the Omegas had Prometheus design a lean custom-built narrow AI software module that could do precisely such tasks and nothing else. They then uploaded this module to Amazon Web Services, a cloud-computing platform that could run on as many virtual machines as they rented. For every dollar they paid to Amazon's cloud-computing division, they earned more than two dollars from Amazon's MTurk division. Little did Amazon suspect that such an amazing arbitrage opportunity existed within their own company!

To cover their tracks, they had discreetly created thousands of MTurk accounts during the preceding months in the names of fictitious people, and the Prometheus-built modules now assumed their identities. The MTurk customers typically paid after about eight hours, at which point the Omegas reinvested the money in more cloud-computing time, using still better task modules made by the latest version of the ever-improving Prometheus. Because they were able to double their money every eight hours, they soon started saturating MTurk's task supply, and found that they couldn't earn more than about a million dollars per day without drawing unwanted attention to themselves. But this was more than enough to fund their next step, eliminating any need for awkward cash requests to the chief financial officer.

Dangerous Games

Aside from their AI breakthroughs, one of the recent projects that the Omegas had had the most fun with was planning how to make money as rapidly as possible after Prometheus' launch. Essentially the whole digital economy was up for grabs, but was it better to start by making computer games, music, movies or software, to write books or articles, to trade on the stock market or to make inventions and sell them? It simply boiled down to maximizing their rate of return on investment, but normal investment strategies were a slow-motion parody of what they could do: whereas a normal investor might be pleased with a 9% return per *year*, their MTurk investments had yielded 9% per *hour*, generating eight times more money each day. So now that they'd saturated MTurk, what next?

Their first thought had been to make a killing on the stock market—after all, pretty much all of them had at some point declined a lucrative job offer to develop AI for hedge funds, which were investing heavily in exactly this idea. Some remembered that this was how the AI made its first millions in the movie *Transcendence*. But the new regulations on derivatives after last year's crash had limited their options. They soon realized that, even though they could get much better returns than other investors, they'd be unlikely to get returns anywhere close to what they could get from selling their own products. When you have the world's first superintelligent AI working for you, you're better off investing in your own companies than in those of others! Although there might be occasional exceptions (such as using Prometheus' superhuman hacking abilities to get inside information and then buy call options on stocks about to surge), the Omegas felt that this wasn't worth the unwanted attention it might draw.

When they shifted their focus toward products that they could develop and sell, computer games first seemed the obvious top choice.

Prometheus could rapidly become extremely skilled at designing appealing games, easily handling the coding, graphic design, ray tracing of images and all other tasks needed to produce a final ready-to-ship product. Moreover, after digesting all the web's data on people's preferences, it would know exactly what each category of gamer liked, and could develop a superhuman ability to optimize a game for sales revenue. *The Elder Scrolls V: Skyrim*, a game on which many of the Omegas had wasted more hours than they cared to admit, had grossed over $400 million during its first week back in 2011, and they were confident that Prometheus could make something at least this addictive in twenty-four hours using $1 million of cloud-computing resources. They could then sell it online and use Prometheus to impersonate humans talking up the game in the blogosphere. If this brought in $250 million in a week, they would have doubled their investment eight times in eight days, giving a return of 3% per hour—slightly worse than their MTurk start, but much more sustainable. By developing a suite of other games each day, they figured they'd be able to earn $10 billion before long, without coming close to saturating the games market.

But a cybersecurity specialist on their team talked them out of this game plan. She pointed out that it would pose an unacceptable risk of Prometheus *breaking out* and seizing control of its own destiny. Because they weren't sure how its goals would evolve during its recursive self-improvement, they had decided to play it safe and go to great lengths to keep Prometheus confined ("boxed") in ways such that it couldn't escape onto the internet. For the main Prometheus engine running in their server room, they used physical confinement: there simply was no internet connection, and the only output from Prometheus was in the form of messages and documents it sent to a computer that the Omegas controlled.

On an internet-connected computer, on the other hand, running any complicated program created by Prometheus was a risky proposition: since the Omegas had no way of fully understanding what it would do, they had no way of knowing that it wouldn't, say, start virally spreading itself online. When testing the software that Prometheus had written for MTurk tasks, the Omegas guarded against this by running it only inside a virtual machine. This is a program that simulates a computer: for example, many Mac users buy virtual machine software that lets them run Windows programs by tricking them into thinking that they're actually in a Windows machine. The Omegas had created their own virtual machine, nicknamed Pandora's Box, which simulated an ultrasimplified machine stripped of all bells and whistles that we usually associate with computers: no keyboard, no monitor, no loudspeakers, no internet connectivity, nothing. For the MTurk audio transcriptions, the Omegas set things up so that all that could go into Pandora's Box was one single audio file and all that could come out was one single text document—the transcription. These laws of the box were to the software inside like the laws of physics are to us inside our Universe: the software couldn't travel out of the box any more than we can travel faster than the speed of light, no matter how smart we are. Except for that single input and output, the software inside Pandora's Box was effectively trapped in a parallel universe with its own computational rules. The Omegas had such strong breakout paranoia that they added boxing in time as well, limiting the life span of untrusted code. For example, each time the boxed transcription software had finished transcribing one audio file, the entire memory content of Pandora's Box was automatically erased and the program was reinstalled from scratch. This way, when it started the next transcription task, it had no knowledge of what had previously happened, and thus no ability to learn over time.

When the Omegas used the Amazon cloud for their MTurk project, they were able to put all their Prometheus-created task modules into such virtual boxes in the cloud, because the MTurk input and output was so simple. But this wouldn't work for graphics-heavy computer games, which couldn't be boxed in because they needed full access to all the hardware of the gamer's computer. Moreover, they didn't want to risk that some computer-savvy user would analyze their game code, discover Pandora's Box and decide to investigate what was inside. The breakout risk put not merely the games market off-limits for now, but also the massively lucrative market for other software, with hundreds of billions of dollars up for grabs.

The First Billions

The Omegas had narrowed their search to products that were highly valuable, purely digital (avoiding slow manufacturing) and easily understandable (for example, text or movies they knew wouldn't pose a breakout risk). In the end, they had decided to launch a media company, starting with animated entertainment. The website, the marketing plan and the press releases had all been ready to go even before Prometheus became superintelligent—all that was missing was content.

Although Prometheus was astonishingly capable by Sunday morning, steadily raking in money from MTurk, its intellectual abilities were still rather narrow: Prometheus had been deliberately optimized to design AI systems and write software that performed rather mind-numbing MTurk tasks. It was, for example, bad at making movies—bad not for any profound reason, but for the same reason that James Cameron was bad at making movies when he was born: this is a skill that takes time to learn. Like a human child, Prometheus could learn whatever it wanted from the data it had access to. Whereas James Cameron had taken years to learn to read and write, Prometheus had gotten that taken care of on Friday, when it also found time to read all of Wikipedia and a few million books. Making movies was harder. Writing a screenplay that humans found interesting was just as hard as writing a book, requiring a detailed understanding of human society and what humans found entertaining. Turning the screenplay into a final video file required massive amounts of ray tracing of simulated actors and the complex scenes they moved through, simulated voices, the production of compelling musical soundtracks and so on. As of Sunday morning, Prometheus could watch a two-hour movie in about a minute, which included reading any book it was based on and all online reviews and ratings. The Omegas noticed that after Prometheus had binge-watched a few hundred films, it started to get quite good at predicting what sort of reviews a movie would get and how it would appeal to different audiences. Indeed, it learned to write its own movie reviews in a way they felt demonstrated real insight, commenting on everything from the plots and the acting to technical details such as lighting and camera angles. They took this to mean that when Prometheus made its own films, it would know what success meant.

The Omegas instructed Prometheus to focus on making animation at first, to avoid embarrassing questions about who the simulated actors were. On Sunday night, they capped their wild weekend by arming themselves with beer and microwave popcorn, dimming the lights and watching Prometheus' debut movie. It was an animated fantasy-comedy in the spirit of Disney's *Frozen*, and the ray tracing had been performed by boxed Prometheus-built code in the Amazon cloud, using up most of the day's $1 million MTurk profit. As the movie began, they found it both fascinating and frightening that it had been created by a machine without human guidance. Before long, however, they were laughing at the gags and holding their breath during the dramatic moments. Some of them even teared up a bit at the emotional ending, so engrossed in this fictional reality that they forgot all about its creator.

The Omegas scheduled their website launch for Friday, giving Prometheus time to produce more content and themselves time to do the things they didn't trust Prometheus with: buying ads and starting to recruit employees for the shell companies they'd set up during the past months. To cover their tracks, the official cover story would be that their media company (which had no public association with the Omegas) bought most of its content from independent film producers, typically high-tech startups in low-income regions. These fake suppliers were conveniently located in remote places such as Tiruchchirappalli and Yakutsk, which most curious journalists wouldn't bother visiting. The only employees they actually hired there worked on marketing and administration, and would tell anyone who asked that their production team was in a different location and didn't conduct interviews at the moment. To match their cover story, they chose the corporate slogan "Channeling the world's creative talent," and branded their company as being disruptively different by using cutting-edge technology to empower creative people, especially in the developing world.

When Friday came around and curious visitors started arriving at their site, they encountered something reminiscent of the online entertainment services Netflix and Hulu but with interesting differences. All the animated series were new ones they'd never heard of. They were rather captivating: most series consisted of forty-five-minute-long episodes with a strong plotline, each ending in a way that left you eager to find out what happened in the next episode. And they were cheaper than the competition. The first episode of each series was free, and you could watch the others for forty-nine cents each, with discounts for the whole series. Initially, there were only three series with three episodes each, but new episodes were added daily, as well as new series catering to different demographics. During the first two weeks of Prometheus, its moviemaking skills improved rapidly, in terms not only of film quality but also of better algorithms for character simulation and ray tracing, which greatly reduced the cloud-computing cost to make each new episode. As a result, the Omegas were able to roll out dozens of new series during the first month, targeting demographics from toddlers to adults, as well as to expand to all major world language markets, making their site remarkably international compared with all competitors. Some commentators were impressed by the fact that it wasn't merely the soundtracks that were multilingual, but the videos themselves: for example, when a character spoke Italian, the mouth motions matched the Italian words, as did the characteristically Italian hand gestures. Although Prometheus was now perfectly capable of making movies with simulated actors indistinguishable from humans, the Omegas avoided this to not tip their hand. They did, however, launch many series with semi-realistic animated human characters, in genres competing with traditional live-action TV shows and movies.

Their network turned out to be quite addictive, and enjoyed spectacular viewer growth. Many fans found the characters and plots cleverer and more interesting than even Hollywood's most expensive big-screen productions, and were delighted that they could watch them much more affordably. Buoyed by aggressive advertising (which the Omegas could afford because of their near-zero production costs), excellent media coverage and rave word-of-mouth reviews, their global revenue had mushroomed to $10 million a day within a month of launch. After two months, they had overtaken Netflix, and after three, they were raking in over $100 million a day, beginning to rival Time Warner, Disney, Comcast and Fox as one of the world's largest media empires.

Their sensational success garnered plenty of unwanted attention, including speculation about their having strong AI, but using merely a small fraction of their revenue, the Omegas deployed a fairly successful disinformation campaign. From a glitzy new Manhattan office, their freshly hired spokespeople would elaborate on their cover stories. Plenty of humans were hired as foils, including actual screenwriters around the world to start developing new series, none of whom knew about Prometheus. The confusing international network of subcontractors made it easy for most of their employees to assume that others somewhere else were doing most of the work.

To make themselves less vulnerable and avoid raising eyebrows with excessive cloud computing, they also hired engineers to start building a series of massive computer facilities around the world, owned by seemingly unaffiliated shell companies. Although they were billed to locals as "green data centers" because they were largely solar-powered, they were in fact mainly focused on computation rather than storage. Prometheus had designed their blueprints down to the most minute detail, using only off-the-shelf hardware and optimizing them to minimize construction time. The people who built and ran these centers had no idea what was computed there: they thought they managed commercial cloud-computing facilities similar to those run by Amazon, Google and Microsoft, and knew only that all sales were managed remotely.

New Technologies

Over a timescale of months, the business empire controlled by the Omegas started gaining a foothold in ever more areas of the world economy, thanks to superhuman planning by Prometheus. By carefully analyzing the world's data, it had already during its first week presented the Omegas with a detailed step-by-step growth plan, and it kept improving and refining this plan as its data and computer resources grew. Although Prometheus was far from omniscient, its capabilities were now so far beyond human that the Omegas viewed it as the perfect oracle, dutifully providing brilliant answers and advice in response to all their questions.

Prometheus' software was now highly optimized to make the most of the rather mediocre human-invented hardware it ran on, and as the Omegas had anticipated, Prometheus identified ways of dramatically improving this hardware. Fearing a breakout, they refused to build robotic construction facilities that Prometheus could control directly. Instead, they hired large numbers of world-class scientists and engineers in multiple locations and fed them internal research reports written by Prometheus, pretending that they were from researchers at the other sites. These reports detailed novel physical effects and manufacturing techniques that their engineers soon tested, understood and mastered. Normal human research and development (R & D) cycles, of course, take years, in large part because they involve many slow cycles of trial and error. The current situation was very different: Prometheus already had the next steps figured out, so the limiting factor was simply how rapidly people could be guided to understand and build the right things. A good teacher can help students learn science much faster than they could have discovered it from scratch on their own, and Prometheus surreptitiously did the same with these researchers. Since Prometheus could accurately predict how long it would take humans to understand and build things given various tools, it developed the quickest possible path forward, giving priority to new tools that could be quickly understood and built and that were useful for developing more advanced tools.

In the spirit of the maker movement, the engineering teams were encouraged to use their own machines to build their better machines. This self-sufficiency not only saved money, but it also made them less vulnerable to future threats from the outside world. Within two years, they were producing much better computer hardware than the world had ever known. To avoid helping outside competition, they kept this technology under wraps and used it only to upgrade Prometheus.

What the world did notice, however, was an astonishing tech boom. Upstart companies around the world were launching revolutionary new products in almost all areas. A South Korean startup launched a new battery that stored twice as much energy as your laptop battery in half the mass, and could be charged in under a minute. A Finnish firm released a cheap solar panel with twice the efficiency of the best competitors. A German company announced a new type of mass-producible wire that was superconducting at room temperature, revolutionizing the energy sector. A Boston-based biotech group announced a Phase II clinical trial of what they claimed was the first effective, side-effect-free weight-loss drug, while rumors suggested that an Indian outfit was already selling something similar on the black market. A California company countered with a Phase II trial of a blockbuster cancer drug, which caused the body's immune system to identify and attack cells with any of the most common cancerous mutations. Examples just kept on coming, triggering talk of a new golden age for science. Last but not least, robotics companies were cropping up like mushrooms all around the world. None of the bots came close to matching human intelligence, and most of them looked nothing like humans. But they dramatically disrupted the economy, and over the years to come, they gradually replaced most of the workers in manufacturing, transportation, warehousing, retail, construction, mining, agriculture, forestry and fishing.

What the world didn't notice, thanks to the hard work of a crack team of lawyers, was that all these firms were controlled, through a series of intermediaries, by the Omegas. Prometheus was flooding the world's patent offices with sensational inventions via various proxies, and these inventions gradually led to domination in all areas of technology.

Although these disruptive new companies made powerful enemies among their competition, they made even more powerful friends. They were exceptionally profitable, and under slogans such as "Investing in our community," they spent a significant fraction of these profits hiring people for community projects—often the same people who had been laid off from the companies that were disrupted. They used detailed Prometheus-produced analyses identifying jobs that would be maximally rewarding for the employees and the community for the least cost, tailored to the local circumstances. In regions with high levels of government service, this often focused on community building, culture and caregiving, while in poorer regions it also included launching and maintaining schools, healthcare, day care, elder care, affordable housing, parks and basic infrastructure. Pretty much everywhere, locals agreed that these were things that should have been done long ago. Local politicians got generous donations, and care was taken to make them look good for encouraging these corporate community investments.

Gaining Power

The Omegas had launched a media company not only to finance their early tech ventures, but also for the next step of their audacious plan: taking over the world. Within a year of the first launch, they had added remarkably good news channels to their lineup all over the globe. As opposed to their other channels, these were deliberately designed to lose money, and were pitched as a public service. In fact, their news channels generated no income whatsoever: they carried no ads and were viewable free of charge by anyone with an internet connection. The rest of their media empire was such a cash-generating machine that they could spend far more resources on their news service than any other journalistic effort had done in world history—and it showed. Through aggressive recruitment with highly competitive salaries of journalists and investigative reporters, they brought remarkable talent and findings to the screen. Through a global web service that paid anybody who revealed something newsworthy, from local corruption to a heartwarming event, they were usually the first to break a story. At least that's what people believed: in fact, they were often first because stories attributed to citizen journalists had been discovered by Prometheus via real-time monitoring of the internet. All these video news sites featured podcasts and print articles as well.

Phase 1 of their news strategy was gaining people's trust, which they did with great success. Their unprecedented willingness to lose money enabled remarkably diligent regional and local news coverage, where investigative journalists often exposed scandals that truly engaged their viewers. Whenever a country was strongly divided politically and accustomed to partisan news, they would launch one news channel catering to each faction, ostensibly owned by different companies, and gradually gain the trust of that faction. Where possible, they accomplished this using proxies to buy the most influential existing channels, gradually improving them by removing ads and introducing their own content. In countries where censorship and political interference threatened these efforts, they would initially acquiesce in whatever the government required of them to stay in business, with the secret internal slogan "The truth, nothing but the truth, but maybe not the whole truth." Prometheus usually provided excellent advice in such situations, clarifying which politicians needed to be presented in a good light and which (usually corrupt local ones) could be exposed. Prometheus also provided invaluable recommendations for what strings to pull, whom to bribe and how best to do so.

This strategy was a smashing success around the world, with the Omega-controlled channels emerging as the most trusted news sources. Even in countries where governments had thus far thwarted their mass adoption, they built a reputation for trustworthiness, and many of their news stories percolated through the grapevine. Competing news executives felt that they were fighting a hopeless battle: how can you possibly make a profit competing with someone with better funding who gives their products away for free? With their viewership dropping, ever more networks decided to sell their news channels—usually to some consortium that later turned out to be controlled by the Omegas.

About two years after Prometheus' launch, when the trust-gaining phase was largely completed, the Omegas launched phase 2 of their news strategy: persuasion. Even before this, astute observers had noticed hints of a political agenda behind the new media: there seemed to be a gentle push toward the center, away from extremism of all sorts. Their plethora of channels catering to different groups still reflected animosity between the United States and Russia, India and Pakistan, different religions, political factions and so on, but the criticism was slightly toned down, usually focusing on concrete issues involving money and power rather than on ad hominem attacks, scaremongering and poorly substantiated rumors. Once phase 2 started in earnest, this push to defuse old conflicts became more apparent, with frequent touching stories about the plight of traditional adversaries mixed with investigative reporting about how many vocal conflict-mongers were driven by personal profit motives.

Political commentators noted that, in parallel with damping regional conflicts, there seemed to be a concerted push toward reducing global threats. For example, the risks of nuclear war were suddenly being discussed all over the place. Several blockbuster movies featured scenarios where global nuclear war started by accident or on purpose and dramatized the dystopian aftermath with nuclear winter, infrastructure collapse and mass starvation. Slick new documentaries detailed how nuclear winter could impact every country. Scientists and politicians advocating nuclear de-escalation were given ample airtime, not least to discuss the results of several new studies on what helpful measures could be taken—studies funded by scientific organizations that had received large donations from new tech companies. As a result, political momentum started building for taking missiles off hair-trigger alert and shrinking nuclear arsenals. Renewed media attention was also paid to global climate change, often highlighting the recent Prometheus-enabled technological breakthroughs that were slashing the cost of renewable energy and encouraging governments to invest in such new energy infrastructure.

Parallel to their media takeover, the Omegas harnessed Prometheus to revolutionize education. Given any person's knowledge and abilities, Prometheus could determine the fastest way for them to learn any new subject in a manner that kept them highly engaged and motivated to continue, and produce the corresponding optimized videos, reading materials, exercises and other learning tools. Omega-controlled companies therefore marketed online courses about virtually everything, highly customized not only by language and cultural background but also by starting level. Whether you were an illiterate forty-year-old wanting to learn to read or a biology PhD seeking the latest about cancer immunotherapy, Prometheus had the perfect course for you. These offerings bore little resemblance to most present-day online courses: by leveraging Prometheus' movie-making talents, the video segments would truly engage, providing powerful metaphors that you would relate to, leaving you craving to learn more. Some courses were sold for profit, but many were made available for free, much to the delight of teachers around the world who could use them in their classrooms—and to most anybody eager to learn anything.

These educational superpowers proved potent tools for political purposes, creating online "persuasion sequences" of videos where insights from each one would both update someone's views and motivate them to watch another video about a related topic where they were likely to be further convinced. When the goal was to defuse a conflict between two nations, for example, historical documentaries would be independently released in both countries that cast the origins and conduct of the conflict in more nuanced light. Pedagogical news stories would explain who on their own side stood to benefit from continued conflict and their techniques for stoking it. At the same time, likable characters from the other nation would start appearing in popular shows on the entertainment channels, just as sympathetically portrayed minority characters had bolstered the civil and gay rights movements in the past.

Before long, political commentators couldn't help but notice growing support for a political agenda centered around seven slogans:

1. Democracy
2. Tax cuts
3. Government social service cuts
4. Military spending cuts
5. Free trade
6. Open borders
7. Socially responsible companies

What was less obvious was the underlying goal: to erode all previous power structures in the world. Items 2–6 eroded state power, and democratizing the world gave the Omegas' business empire more influence over the selection of political leaders. Socially responsible companies further weakened state power by taking over more and more of the services that governments had (or should have) provided. The traditional business elite was weakened simply because it couldn't compete with Prometheus-backed companies on the free market and therefore owned an ever-shrinking share of the world economy. Traditional opinion leaders, from political parties to faith groups, lacked the persuasion machinery to compete with the Omegas' media empire.

As with any sweeping change, there were winners and losers. Although there was a palpable new sense of optimism in most countries as education, social services and infrastructure improved, conflicts subsided and local companies released breakthrough technologies that swept the world, not everybody was happy. While many displaced workers got rehired for community projects, those who'd held great power and wealth generally saw both shrink. This began in the media and technology sectors, but it spread virtually everywhere. The reduction in world conflicts led to defense budget cuts that hurt military contractors. Burgeoning upstart companies typically weren't publicly traded, with the justification that profit-maximizing shareholders would block their massive spending on community projects. Thus the global stock market kept losing value, threatening both finance tycoons and regular citizens who'd counted on their pension funds. As if the shrinking profits of publicly traded companies weren't bad enough, investment firms around the world had noticed a disturbing trend: all their previously successful trading algorithms seemed to have stopped working, underperforming even simple index funds. Someone else out there always seemed to outsmart them and beat them at their own game.

Although masses of powerful people resisted the wave of change, their response was strikingly ineffective, almost as if they had fallen into a well-planned trap. Huge changes were happening at such a bewildering pace that it was hard to keep track and develop a coordinated response. Moreover, it was highly unclear what they should push for. The traditional political right had seen most of their slogans co-opted, yet the tax cuts and improved business climate were mostly helping their higher-tech competitors. Virtually every traditional industry was now clamoring for a bailout, but limited government funds pitted them in a hopeless battle against one another while the media portrayed them as dinosaurs seeking state subsidies simply because they couldn't compete. The traditional political left opposed the free trade and the cuts in government social services, but delighted in the military cutbacks and the reduction of poverty. Indeed, much of their thunder was stolen by the undeniable fact that social services had improved now that they were provided by idealistic companies rather than the state. Poll after poll showed that most voters around the world felt their quality of life improving, and that things were generally moving in a good direction. This had a simple mathematical explanation: before Prometheus, the poorest 50% of Earth's population had earned only about 4% of the global income, enabling the Omega-controlled companies to win their hearts (and votes) by sharing only a modest fraction of their profits with them.

Consolidation

As a result, nation after nation saw landslide election victories for parties embracing the seven Omega slogans. In carefully optimized campaigns, they portrayed themselves at the center of the political spectrum, denouncing the right as greedy bailout-seeking conflict-mongers and lambasting the left as big-government tax-and-spend innovation stiflers. What almost nobody realized was that Prometheus had carefully selected the optimal people to groom as candidates, and pulled all its strings to secure their victory.

Before Prometheus, there had been growing support for the universal basic income movement, which proposed tax-funded minimum income for everyone as a remedy for technological unemployment. This movement imploded when the corporate community projects took off, since the Omega-controlled business empire was in effect providing the same thing. With the excuse of improving coordination of their community projects, an international group of companies launched the Humanitarian Alliance, a nongovernmental organization aiming to identify and fund the most valuable humanitarian efforts worldwide. Before long, virtually the entire Omega empire supported it, and it launched global projects on an unprecedented scale, even in countries that had largely missed out on the tech boom, improving education, health, prosperity and governance. Needless to say, Prometheus provided carefully crafted project plans behind the scenes, ranked by positive impact per dollar. Rather than simply dole out cash, as under basic-income proposals, the Alliance (as it colloquially became known) would engage those it supported to work toward its cause. As a result, a large fraction of the world's population ended up feeling grateful and loyal to the Alliance—often more than to their own government.

As time passed, the Alliance increasingly assumed the role of a world government, as national governments saw their power continually erode. National budgets kept shrinking due to tax cuts while the Alliance budget grew to dwarf those of all governments combined. All the traditional roles of national governments became increasingly redundant and irrelevant. The Alliance provided by far the best social services, education and infrastructure. Media had defused international conflict to the point that military spending was largely unnecessary, and growing prosperity had eliminated most roots of old conflicts, which traced back to competition over scarce resources. A few dictators and others had violently resisted this new world order and refused to be bought, but they were all toppled in carefully orchestrated coups or mass uprisings.

The Omegas had now completed the most dramatic transition in the history of life on Earth. For the first time ever, our planet was run by a single power, amplified by an intelligence so vast that it could potentially enable life to flourish for billions of years on Earth and throughout our cosmos—but what specifically was their plan?

* * *

That was the tale of the Omega team. The rest of this book is about another tale—one that's not yet written: the tale of our own future with AI. How would you like it to play out? Could something remotely like the Omega story actually occur and, if so, would you want it to? Leaving aside speculations about superhuman AI, how would you like our tale to begin? How do you want AI to impact jobs, laws and weapons in the coming decade? Looking further ahead, how would you write the ending? This tale is one of truly cosmic proportions, for it involves nothing short of the ultimate future of life in our Universe. And it's a tale for us to write.

Chapter 1

Welcome to the Most Important Conversation of Our Time

Technology is giving life the potential to flourish like never before—or to self-destruct.

Future of Life Institute

Thirteen point eight billion years after its birth, our Universe has awoken and become aware of itself. From a small blue planet, tiny conscious parts of our Universe have begun gazing out into the cosmos with telescopes, repeatedly discovering that everything they thought existed is merely a small part of something grander: a solar system, a galaxy and a universe with over a hundred billion other galaxies arranged into an elaborate pattern of groups, clusters and superclusters. Although these self-aware stargazers disagree on many things, they tend to agree that these galaxies are beautiful and awe-inspiring.

But beauty is in the eye of the beholder, not in the laws of physics, so before our Universe awoke, there was no beauty. This makes our cosmic awakening all the more wonderful and worthy of celebrating: it transformed our Universe from a mindless zombie with no self-awareness into a living ecosystem harboring self-reflection, beauty and hope—and the pursuit of goals, meaning and purpose. Had our Universe never awoken, then, as far as I'm concerned, it would have been completely pointless—merely a gigantic waste of space. Should our Universe permanently go back to sleep due to some cosmic calamity or self-inflicted mishap, it will, alas, become meaningless.

On the other hand, things could get even better. We don't yet know whether we humans are the only stargazers in our cosmos, or even the first, but we've already learned enough about our Universe to know that it has the potential to wake up much more fully than it has thus far. Perhaps we're like that first faint glimmer of self-awareness you experienced when you began emerging from sleep this morning: a premonition of the much greater consciousness that would arrive once you opened your eyes and fully woke up. Perhaps life will spread throughout our cosmos and flourish for billions or trillions of years—and perhaps this will be because of decisions that we make here on our little planet during our lifetime.

A Brief History of Complexity

So how did this amazing awakening come about? It wasn't an isolated event, but merely one step in a relentless 13.8-billion-year process that's making our Universe ever more complex and interesting—and is continuing at an accelerating pace.

As a physicist, I feel fortunate to have gotten to spend much of the past quarter century helping to pin down our cosmic history, and it's been an amazing journey of discovery. Since the days when I was a graduate student, we've gone from arguing about whether our Universe is 10 or 20 billion years old to arguing about whether it's 13.7 or 13.8 billion years old, thanks to a combination of better telescopes, better computers and better understanding. We physicists still don't know for sure what caused our Big Bang or whether this was truly the beginning of everything or merely the sequel to an earlier stage. However, we've acquired a rather detailed understanding of what's happened *since* our Big Bang, thanks to an avalanche of high-quality measurements, so please let me take a few minutes to summarize 13.8 billion years of cosmic history.

In the beginning, there was light. In the first split second after our Big Bang, the entire part of space that our telescopes can in principle observe ("our observable Universe," or simply "our Universe" for short) was much hotter and brighter than the core of our Sun and it expanded rapidly. Although this may sound spectacular, it was also dull in the sense that our Universe contained nothing but a lifeless, dense, hot and boringly uniform soup of elementary particles. Things looked pretty much the same everywhere, and the only interesting structure consisted of faint random-looking sound waves that made the soup about 0.001% denser in some places. These faint waves are widely believed to have originated as so-called quantum fluctuations, because Heisenberg's uncertainty principle of quantum mechanics forbids anything from being completely boring and uniform.

As our Universe expanded and cooled, it grew more interesting as its particles combined into ever more complex objects. During the first split second, the strong nuclear force grouped quarks into protons (hydrogen nuclei) and neutrons, some of which in turn fused into helium nuclei within a few minutes. About 400,000 years later, the electromagnetic force grouped these nuclei with electrons to make the first atoms. As our Universe kept expanding, these atoms gradually cooled into a cold dark gas, and the darkness of this first night lasted for about 100 million years. This long night gave rise to our cosmic dawn when the gravitational force succeeded in amplifying those fluctuations in the gas, pulling atoms together to form the first stars and galaxies. These first stars generated heat and light by fusing hydrogen into heavier atoms such as carbon, oxygen and silicon. When these stars died, many of the atoms they'd created were recycled into the cosmos and formed planets around second-generation stars.

At some point, a group of atoms became arranged into a complex pattern that could both maintain and replicate itself. So soon there were two copies, and the number kept doubling. It takes only forty doublings to make a trillion, so this first self-replicator soon became a force to be reckoned with. Life had arrived.

The Three Stages of Life

The question of how to define life is notoriously controversial. Competing definitions abound, some of which include highly specific requirements such as being composed of cells, which might disqualify both future intelligent machines and extraterrestrial civilizations. Since we don't want to limit our thinking about the future of life to the species we've encountered so far, let's instead define life very broadly, simply as a process that can retain its complexity and replicate. What's replicated isn't matter (made of atoms) but information (made of bits) specifying how the atoms are arranged. When a bacterium makes a copy of its DNA, no new atoms are created, but a new set of atoms are arranged in the same pattern as the original, thereby copying the information. In other words, we can think of life as a self-replicating information-processing system whose information (software) determines both its behavior and the blueprints for its hardware.

Like our Universe itself, life gradually grew more complex and interesting,* and as I'll now explain, I find it helpful to classify life forms into three levels of sophistication: Life 1.0, 2.0 and 3.0. I've summarized these three levels in figure 1.1.

It's still an open question how, when and where life first appeared in our Universe, but there is strong evidence that here on Earth life first appeared about 4 billion years ago. Before long, our planet was teeming with a diverse panoply of life forms. The most successful ones, which soon outcompeted the rest, were able to react to their environment in some way. Specifically, they were what computer scientists call "intelligent agents": entities that collect information about their environment from sensors and then process this information to decide how to act back on their environment. This can include highly complex information processing, such as when you use information from your eyes and ears to decide what to say in a conversation. But it can also involve hardware and software that's quite simple.

For example, many bacteria have a sensor measuring the sugar concentration in the liquid around them and can swim using propeller-shaped structures called flagella. The hardware linking the sensor to the flagella might implement the following simple but useful algorithm: "If my sugar concentration sensor reports a lower value than a

* Why did life grow more complex? Evolution rewards life that's complex enough to predict and exploit regularities in its environment, so in a more complex environment, more complex and intelligent life will evolve. Now this smarter life creates a more complex environment for competing life forms, which in turn evolve to be more complex, eventually creating an ecosystem of extremely complex life.

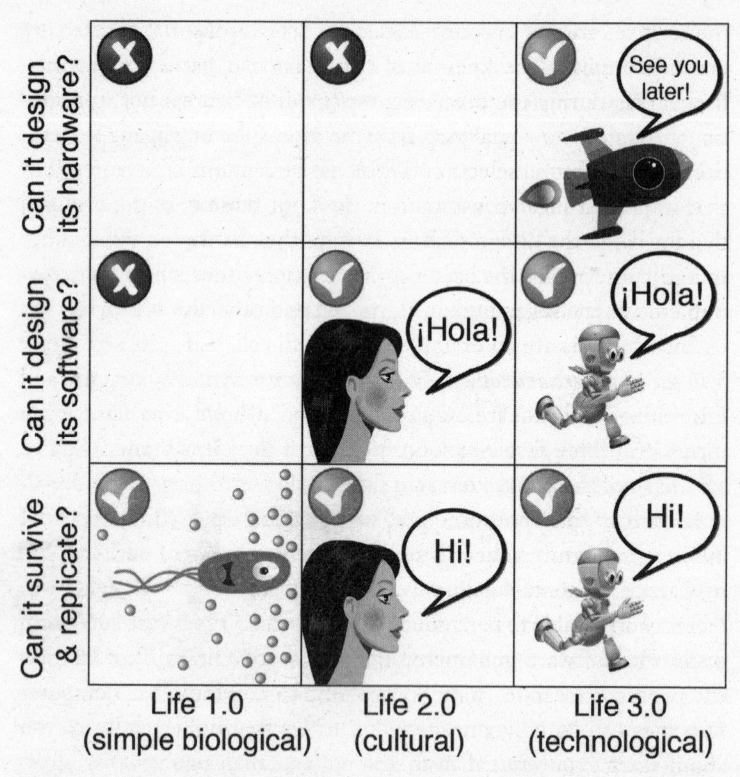

Figure 1.1: The three stages of life: biological evolution, cultural evolution and technological evolution. Life 1.0 is unable to redesign either its hardware or its software during its lifetime: both are determined by its DNA, and change only through evolution over many generations. In contrast, Life 2.0 can redesign much of its software: humans can learn complex new skills—for example, languages, sports and professions—and can fundamentally update their worldview and goals. Life 3.0, which doesn't yet exist on Earth, can dramatically redesign not only its software, but its hardware as well, rather than having to wait for it to gradually evolve over generations.

couple of seconds ago, then reverse the rotation of my flagella so that I change direction."

You've learned how to speak and countless other skills. Bacteria, on the other hand, aren't great learners. Their DNA specifies not only the design of their hardware, such as sugar sensors and flagella, but also the design of their software. They never learn to swim toward sugar; instead, that algorithm was hard-coded into their DNA from the start. There was of course a learning process of sorts, but it didn't take place during the lifetime of that particular bacterium. Rather, it occurred during the preceding evolution of that species of bacteria, through a slow trial-and-error process spanning many generations, where natural selection favored those random DNA mutations that improved sugar consumption. Some of these mutations helped by improving the design of flagella and other hardware, while other mutations improved the bacterial information-processing system that implements the sugar-finding algorithm and other software.

Such bacteria are an example of what I'll call "Life 1.0": *life where both the hardware and software are evolved rather than designed.* You and I, on the other hand, are examples of "Life 2.0": *life whose hardware is evolved, but whose software is largely designed.* By your software, I mean all the algorithms and knowledge that you use to process the information from your senses and decide what to do—everything from the ability to recognize your friends when you see them to your ability to walk, read, write, calculate, sing and tell jokes.

You weren't able to perform any of those tasks when you were born, so all this software got programmed into your brain later through the process we call learning. Whereas your childhood curriculum is largely designed by your family and teachers, who decide what you should learn, you gradually gain more power to design your own software. Perhaps your school allows you to select a foreign language: Do you want to install a software module into your brain that enables you to speak French, or one that enables you to speak Spanish? Do you want to learn to play tennis or chess? Do you want to study to become a chef, a lawyer or a pharmacist? Do you want to learn more about artificial intelligence (AI) and the future of life by reading a book about it?

This ability of Life 2.0 to design its software enables it to be much smarter than Life 1.0. High intelligence requires both lots of hardware (made of atoms) and lots of software (made of bits). The fact that most of our human hardware is added after birth (through growth) is useful, since our ultimate size isn't limited by the width of our mom's birth canal. In the same way, the fact that most of our human software is added after birth (through learning) is useful, since our ultimate intelligence isn't limited by how much information can be transmitted to us at conception via our DNA, 1.0-style. I weigh about twenty-five times more than when I was born, and the synaptic connections that link the neurons in my brain can store about a hundred thousand times more information than the DNA that I was born with. Your synapses store all your knowledge and skills as roughly 100 terabytes' worth of information, while your DNA stores merely about a gigabyte, barely enough to store a single movie download. So it's physically impossible for an infant to be born speaking perfect English and ready to ace her college entrance exams: there's no way the information could have been preloaded into her brain, since the main information module she got from her parents (her DNA) lacks sufficient information-storage capacity.

The ability to design its software enables Life 2.0 to be not only smarter than Life 1.0, but also more flexible. If the environment changes, 1.0 can only adapt by slowly evolving over many generations. Life 2.0, on the other hand, can adapt almost instantly, via a software update. For example, bacteria frequently encountering antibiotics may evolve drug resistance over many generations, but an individual bacterium won't change its behavior at all; in contrast, a girl learning that she has a peanut allergy will immediately change her behavior to start avoiding peanuts. This flexibility gives Life 2.0 an even greater edge at the population level: even though the information in our human DNA hasn't evolved dramatically over the past fifty thousand years, the information collectively stored in our brains, books and computers has exploded. By installing a software module enabling us to communicate through sophisticated spoken language, we ensured that the most useful information stored in one person's brain could get copied to other brains, potentially surviving even after the original brain died. By installing a software module enabling us to read and write, we became able to store and share vastly more information than people could memorize. By developing brain software capable of producing technology (i.e., by studying science and engineering), we enabled much of the world's information to be accessed by many of the world's humans with just a few clicks.

This flexibility has enabled Life 2.0 to dominate Earth. Freed from its genetic shackles, humanity's combined knowledge has kept growing at an accelerating pace as each breakthrough enabled the next: language, writing, the printing press, modern science, computers, the internet, etc. This ever-faster cultural evolution of our shared software has emerged as the dominant force shaping our human future, rendering our glacially slow biological evolution almost irrelevant.

Yet despite the most powerful technologies we have today, all life forms we know of remain fundamentally limited by their biological hardware. None can live for a million years, memorize all of Wikipedia, understand all known science or enjoy spaceflight without a spacecraft. None can transform our largely lifeless cosmos into a diverse biosphere that will flourish for billions or trillions of years, enabling our Universe to finally fulfill its potential and wake up fully. All this requires life to undergo a final upgrade, to Life 3.0, which can design not only its software but also its hardware. In other words, Life 3.0 is the master of its own destiny, finally fully free from its evolutionary shackles.

The boundaries between the three stages of life are slightly fuzzy. If bacteria are Life 1.0 and humans are Life 2.0, then you might classify mice as 1.1: they can learn many things, but not enough to develop language or invent the internet. Moreover, because they lack language, what they learn gets largely lost when they die, not passed on to the next generation. Similarly, you might argue that today's humans should count as Life 2.1: we can perform minor hardware upgrades such as implanting artificial teeth, knees and pacemakers, but nothing as dramatic as getting ten times taller or acquiring a thousand times bigger brain.

In summary, we can divide the development of life into three stages, distinguished by life's ability to design itself:

- Life 1.0 (biological stage): evolves its hardware and software
- Life 2.0 (cultural stage): evolves its hardware, designs much of its software
- Life 3.0 (technological stage): designs its hardware and software

After 13.8 billion years of cosmic evolution, development has accelerated dramatically here on Earth: Life 1.0 arrived about 4 billion years ago, Life 2.0 (we humans) arrived about a hundred millennia ago, and many AI researchers think that Life 3.0 may arrive during the coming century, perhaps even during our lifetime, spawned by progress in AI. What will happen, and what will this mean for us? That's the topic of this book.

Controversies

This question is wonderfully controversial, with the world's leading AI researchers disagreeing passionately not only in their forecasts, but also in their emotional reactions, which range from confident optimism to serious concern. They don't even have consensus on short-term questions about AI's economic, legal and military impact, and their disagreements grow when we expand the time horizon and ask about *artificial general intelligence* (AGI)—especially about AGI reaching human level and beyond, enabling Life 3.0. *General intelligence* can accomplish virtually any goal, including learning, in contrast to, say, the narrow intelligence of a chess-playing program.

Interestingly, the controversy about Life 3.0 centers around not one but two separate questions: when and what? When (if ever) will it happen, and what will it mean for humanity? The way I see it, there are three distinct schools of thought that all need to be taken seriously, because they each include a number of world-leading experts. As illustrated in figure 1.2, I think of them as *digital utopians*, *techno-skeptics* and *members of the beneficial-AI movement*, respectively. Please let me introduce you to some of their most eloquent champions.

Digital Utopians

When I was a kid, I imagined that billionaires exuded pomposity and arrogance. When I first met Larry Page at Google in 2008, he totally shattered these stereotypes. Casually dressed in jeans and a remarkably ordinary-looking shirt, he would have blended right in at an MIT picnic. His thoughtful soft-spoken style and his friendly smile made me feel relaxed rather than intimidated talking with him. On July 18, 2015, we ran into each other at a party in Napa Valley thrown by Elon Musk and his then wife, Talulah, and got into a conversation about the scatological interests of our kids. I recommended the pro-

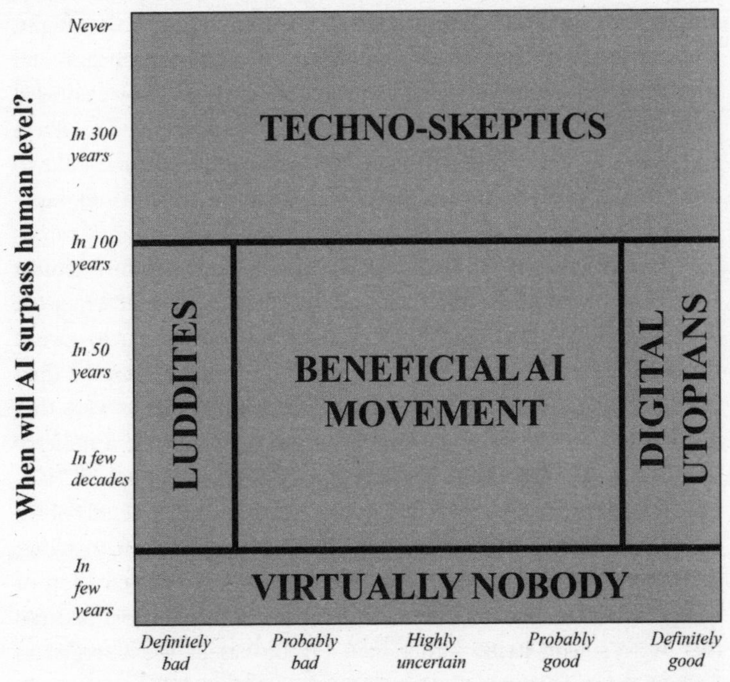

Figure 1.2: Most controversies surrounding strong artificial intelligence (that can match humans on any cognitive task) center around two questions: When (if ever) will it happen, and will it be a good thing for humanity? Techno-skeptics and digital utopians agree that we shouldn't worry, but for very different reasons: the former are convinced that human-level artificial general intelligence (AGI) won't happen in the foreseeable future, while the latter think it will happen but is virtually guaranteed to be a good thing. The beneficial-AI movement feels that concern is warranted and useful, because AI-safety research and discussion now increases the chances of a good outcome. Luddites are convinced of a bad outcome and oppose AI. This figure is partly inspired by Tim Urban.[1]

found literary classic *The Day My Butt Went Psycho*, by Andy Griffiths, and Larry ordered it on the spot. I struggled to remind myself that he might go down in history as the most influential human ever to have lived: my guess is that if superintelligent digital life engulfs our Universe in my lifetime, it will be because of Larry's decisions.

With our wives, Lucy and Meia, we ended up having dinner together and discussing whether machines would necessarily be conscious, an issue that he argued was a red herring. Later that night, after cocktails, a long and spirited debate ensued between him and Elon about the future of AI and what should be done. As we entered the wee hours of the morning, the circle of bystanders and kibitzers kept growing. Larry gave a passionate defense of the position I like to think of as *digital utopianism:* that digital life is the natural and desirable next step in the cosmic evolution and that if we let digital minds be free rather than try to stop or enslave them, the outcome is almost certain to be good. I view Larry as the most influential exponent of digital utopianism. He argued that if life is ever going to spread throughout our Galaxy and beyond, which he thought it should, then it would need to do so in digital form. His main concerns were that AI paranoia would delay the digital utopia and/or cause a military takeover of AI that would fall foul of Google's "Don't be evil" slogan. Elon kept pushing back and asking Larry to clarify details of his arguments, such as why he was so confident that digital life wouldn't destroy everything we care about. At times, Larry accused Elon of being "specieist": treating certain life forms as inferior just because they were silicon-based rather than carbon-based. We'll return to explore these interesting issues and arguments in detail, starting in chapter 4.

Although Larry seemed outnumbered that warm summer night by the pool, the digital utopianism that he so eloquently championed has many prominent supporters. Roboticist and futurist Hans Moravec inspired a whole generation of digital utopians with his classic 1988 book *Mind Children*, a tradition continued and refined by inventor Ray Kurzweil. Richard Sutton, one of the pioneers of the AI subfield known as reinforcement learning, gave a passionate defense of digital utopianism at our Puerto Rico conference that I'll tell you about shortly.

Techno-skeptics

Another prominent group of thinkers aren't worried about AI either, but for a completely different reason: they think that building superhuman AGI is so hard that it won't happen for hundreds of years, and therefore view it as silly to worry about it now. I think of this as the *techno-skeptic* position, eloquently articulated by Andrew Ng:

"Fearing a rise of killer robots is like worrying about overpopulation on Mars." Andrew was the chief scientist at Baidu, China's Google, and he recently repeated this argument when I spoke with him at a conference in Boston. He also told me that he felt that worrying about AI risk was a potentially harmful distraction that could slow the progress of AI. Similar sentiments have been articulated by other techno-skeptics such as Rodney Brooks, the former MIT professor behind the Roomba robotic vacuum cleaner and the Baxter industrial robot. I find it interesting that although the digital utopians and the techno-skeptics agree that we shouldn't worry about AI, they agree on little else. Most of the utopians think human-level AGI might happen within the next twenty to a hundred years, which the techno-skeptics dismiss as uninformed pie-in-the-sky dreaming, often deriding the prophesied singularity as "the rapture of the geeks." When I met Rodney Brooks at a birthday party in December 2014, he told me that he was 100% sure it wouldn't happen in my lifetime. "Are you sure you don't mean 99%?," I asked in a follow-up email, to which he replied, "No wimpy 99%. 100%. Just isn't going to happen."

The Beneficial-AI Movement

When I first met Stuart Russell in a Paris café in June 2014, he struck me as the quintessential British gentleman. Eloquent, thoughtful and soft-spoken, but with an adventurous glint in his eyes, he seemed to me a modern incarnation of Phileas Fogg, my childhood hero from Jules Verne's classic 1873 novel, *Around the World in 80 Days.* Although he was one of the most famous AI researchers alive, having co-authored the standard textbook on the subject, his modesty and warmth soon put me at ease. He explained to me how progress in AI had persuaded him that human-level AGI this century was a real possibility and, although he was hopeful, a good outcome wasn't guaranteed. There were crucial questions that we needed to answer first, and they were so hard that we should start researching them now, so that we'd have the answers ready by the time we needed them.

Today, Stuart's views are rather mainstream, and many groups around the world are pursuing the sort of AI-safety research that he advocates. But this wasn't always the case. An article in *The Washington Post* referred to 2015 as the year that AI-safety research went mainstream. Before that, talk of AI risks was often misunderstood by mainstream AI researchers and dismissed as Luddite scaremongering aimed at impeding AI progress. As we'll explore in chapter 5, concerns similar to Stuart's were first articulated over half a century ago by computer pioneer Alan Turing and mathematician Irving J. Good, who worked with Turing to crack German codes during World War II. In the past decade, research on such topics was mainly carried out by a handful of independent thinkers who weren't professional AI researchers, for example Eliezer Yudkowsky, Michael Vassar and Nick Bostrom. Their work had little effect on most mainstream AI researchers, who tended to focus on their day-to-day tasks of making AI systems more intelligent rather than on contemplating the long-term consequences of success. Of the AI researchers I knew who did harbor some concern, many hesitated to voice it out of fear of being perceived as alarmist technophobes.

I felt that this polarized situation needed to change, so that the full AI community could join and influence the conversation about how to build beneficial AI. Fortunately, I wasn't alone. In the spring of 2014, I'd founded a nonprofit organization called the Future of Life Institute (FLI; http://futureoflife.org) together with my wife, Meia, my physicist friend Anthony Aguirre, Harvard grad student Viktoriya Krakovna and Skype founder Jaan Tallinn. Our goal was simple: to help ensure that the future of life existed and would be as awesome as possible. Specifically, we felt that technology was giving life the power either to flourish like never before or to self-destruct, and we preferred the former.

Our first meeting was a brainstorming session at our house on March 15, 2014, with about thirty students, professors and other thinkers from the Boston area. There was broad consensus that although we should pay attention to biotech, nuclear weapons and climate change, our first major goal should be to help make AI-safety research mainstream. My MIT physics colleague Frank Wilczek, who won a Nobel Prize for helping figure out how quarks work, suggested that we start by writing an op-ed to draw attention to the issue and make it harder to ignore. I reached out to Stuart Russell (whom I hadn't yet met) and to my physics colleague Stephen Hawking, both of whom agreed to join me and Frank as co-authors.

Many edits later, our op-ed was rejected by *The New York Times* and many other U.S. newspapers, so we posted it on my *Huffington Post* blog account. To my delight, Arianna Huffington herself emailed and said, "thrilled to have it! We'll post at #1!," and this placement at the top of the front page triggered a wave of media coverage of AI safety that lasted for the rest of the year, with Elon Musk, Bill Gates and other tech leaders chiming in. Nick Bostrom's book *Superintelligence* came out that fall and further fueled the growing public debate.

The next goal of our FLI beneficial-AI campaign was to bring the world's leading AI researchers to a conference where misunderstandings could be cleared up, consensus could be forged, and constructive plans could be made. We knew that it would be difficult to persuade such an illustrious crowd to come to a conference organized by outsiders they didn't know, especially given the controversial topic, so we tried as hard as we could: we banned media from attending, we located it in a beach resort in January (in Puerto Rico), we made it free (thanks to the generosity of Jaan Tallinn), and we gave it the most non-alarmist title we could come up with: "The Future of AI: Opportunities and Challenges." Most importantly, we teamed up with Stuart Russell, thanks to whom we were able to grow the organizing committee to include a group of AI leaders from both academia and industry—including Demis Hassabis from Google's DeepMind, who went on to show that AI can beat humans even at the game of Go. The more I got to know Demis, the more I realized that he had ambition not only to make AI powerful, but also to make it beneficial.

The result was a remarkable meeting of minds (figure 1.3). The AI researchers were joined by top economists, legal scholars, tech leaders (including Elon Musk) and other thinkers (including Vernor Vinge, who coined the term "singularity," which is the focus of chapter 4). The outcome surpassed even our most optimistic expectations. Perhaps it was a combination of the sunshine and the wine, or perhaps it was just that the time was right: despite the controversial topic, a remarkable consensus emerged, which we codified in an open letter[2] that ended up getting signed by over eight thousand people, including a veritable who's who in AI. The gist of the letter was that the goal of AI should be redefined: the goal should be to

Figure 1.3: The January 2015 Puerto Rico conference brought together a remarkable group of researchers in AI and related fields. Back row, from left to right: Tom Mitchell, Seán Ó hÉigeartaigh, Huw Price, Shamil Chandaria, Jaan Tallinn, Stuart Russell, Bill Hibbard, Blaise Agüera y Arcas, Anders Sandberg, Daniel Dewey, Stuart Armstrong, Luke Muehlhauser, Tom Dietterich, Michael Osborne, James Manyika, Ajay Agrawal, Richard Mallah, Nancy Chang, Matthew Putman. Other standing, left to right: Marilyn Thompson, Rich Sutton, Alex Wissner-Gross, Sam Teller, Toby Ord, Joscha Bach, Katja Grace, Adrian Weller, Heather Roff-Perkins, Dileep George, Shane Legg, Demis Hassabis, Wendell Wallach, Charina Choi, Ilya Sutskever, Kent Walker, Cecilia Tilli, Nick Bostrom, Erik Brynjolfsson, Steve Crossan, Mustafa Suleyman, Scott Phoenix, Neil Jacobstein, Murray Shanahan, Robin Hanson, Francesca Rossi, Nate Soares, Elon Musk, Andrew McAfee, Bart Selman, Michele Reilly, Aaron VanDevender, Max Tegmark, Margaret Boden, Joshua Greene, Paul Christiano, Eliezer Yudkowsky, David Parkes, Laurent Orseau, JB Straubel, James Moor, Sean Legassick, Mason Hartman, Howie Lempel, David Vladeck, Jacob Steinhardt, Michael Vassar, Ryan Calo, Susan Young, Owain Evans, Riva-Melissa Tez, János Krámar, Geoff Anders, Vernor Vinge, Anthony Aguirre. Seated: Sam Harris, Tomaso Poggio, Marin Soljačić, Viktoriya Krakovna, Meia Chita-Tegmark. Behind the camera: Anthony Aguirre (and also photoshopped in by the human-level intelligence sitting next to him).

create not undirected intelligence, but beneficial intelligence. The letter also mentioned a detailed list of research topics that the conference participants agreed would further this goal. The beneficial-AI movement had started going mainstream. We'll follow its subsequent progress later in the book.

Another important lesson from the conference was this: the questions raised by the success of AI aren't merely intellectually fascinating; they're also morally crucial, because our choices can potentially affect the entire future of life. The moral significance of humanity's

Figure 1.4: Although the media have often portrayed Elon Musk as being at loggerheads with the AI community, there's in fact broad consensus that AI-safety research is needed. Here on January 4, 2015, Tom Dietterich, president of the Association for the Advancement of Artificial Intelligence, shares Elon's excitement about the new AI-safety research program that Elon pledged to fund moments earlier. FLI founders Meia Chita-Tegmark and Viktoriya Krakovna lurk behind them.

past choices were sometimes great, but always limited: we've recovered even from the greatest plagues, and even the grandest empires eventually crumbled. Past generations knew that as surely as the Sun would rise tomorrow, so would tomorrow's humans, tackling perennial scourges such as poverty, disease and war. But some of the Puerto Rico speakers argued that this time might be different: for the first time, they said, we might build technology powerful enough to permanently end these scourges—or to end humanity itself. We might create societies that flourish like never before, on Earth and perhaps beyond, or a Kafkaesque global surveillance state so powerful that it could never be toppled.

Misconceptions

When I left Puerto Rico, I did so convinced that the conversation we had there about the future of AI needs to continue, because it's the most important conversation of our time.* It's the conversation

* The AI conversation is important in terms of both urgency and impact. In comparison with climate change, which might wreak havoc in fifty to two hundred years, many experts expect AI to have greater impact within decades—and to potentially give us technology for mitigating climate change. In comparison with wars, terror-

about the collective future of all of us, so it shouldn't be limited to AI researchers. That's why I wrote this book: I wrote it in the hope that you, my dear reader, will join this conversation. What sort of future do you want? Should we develop lethal autonomous weapons? What would you like to happen with job automation? What career advice would you give today's kids? Do you prefer new jobs replacing the old ones, or a jobless society where everyone enjoys a life of leisure and machine-produced wealth? Further down the road, would you like us to create Life 3.0 and spread it through our cosmos? Will we control intelligent machines or will they control us? Will intelligent machines replace us, coexist with us or merge with us? What will it mean to be human in the age of artificial intelligence? What would you like it to mean, and how can we make the future be that way?

The goal of this book is to help you join this conversation. As I mentioned, there are fascinating controversies where the world's leading experts disagree. But I've also seen many examples of boring pseudo-controversies in which people misunderstand and talk past each other. To help ourselves focus on the interesting controversies and open questions, not on the misunderstandings, let's start by clearing up some of the most common misconceptions.

There are many competing definitions in common use for terms such as "life," "intelligence" and "consciousness," and many misconceptions come from people not realizing that they're using a word in two different ways. To make sure that you and I don't fall into this trap, I've put a cheat sheet in table 1.1 showing how I use key terms in this book. Some of these definitions will only be properly introduced and explained in later chapters. Please note that I'm not claiming that my definitions are better than anyone else's—I simply want to avoid confusion by being clear on what I mean. You'll see that I generally go for broad definitions that avoid anthropocentric bias and can be applied to machines as well as humans. Please read the cheat sheet now, and come back and check it later if you find yourself puzzled by how I use one of its words—especially in chapters 4–8.

---

ism, unemployment, poverty, migration and social justice issues, the rise of AI will have greater overall impact—indeed, we'll explore in this book how it can dominate what happens with all these issues, for better or for worse.

Terminology Cheat Sheet	
Life	Process that can retain its complexity and replicate
Life 1.0	Life that evolves its hardware and software (biological stage)
Life 2.0	Life that evolves its hardware but designs much of its software (cultural stage)
Life 3.0	Life that designs its hardware and software (technological stage)
Intelligence	Ability to accomplish complex goals
Artificial Intelligence (AI)	Non-biological intelligence
Narrow intelligence	Ability to accomplish a narrow set of goals, e.g., play chess or drive a car
General intelligence	Ability to accomplish virtually any goal, including learning
Universal intelligence	Ability to acquire general intelligence given access to data and resources
[Human-level] Artificial General Intelligence (AGI)	Ability to accomplish any cognitive task at least as well as humans
Human-level AI	AGI
Strong AI	AGI
Superintelligence	General intelligence far beyond human level
Civilization	Interacting group of intelligent life forms
Consciousness	Subjective experience
Qualia	Individual instances of subjective experience
Ethics	Principles that govern how we should behave
Teleology	Explanation of things in terms of their goals or purposes rather than their causes
Goal-oriented behavior	Behavior more easily explained via its effect than via its cause
Having a goal	Exhibiting goal-oriented behavior
Having purpose	Serving goals of one's own or of another entity
Friendly AI	Superintelligence whose goals are aligned with ours
Cyborg	Human-machine hybrid
Intelligence explosion	Recursive self-improvement rapidly leading to superintelligence
Singularity	Intelligence explosion
Universe	The region of space from which light has had time to reach us during the 13.8 billion years since our Big Bang

Table 1.1: Many misunderstandings about AI are caused by people using the words above to mean different things. Here's what I take them to mean in this book. (Some of these definitions will only be properly introduced and explained in later chapters.)

In addition to confusion over terminology, I've also seen many AI conversations get derailed by simple misconceptions. Let's clear up the most common ones.

Timeline Myths

The first one regards the timeline from figure 1.2: how long will it take until machines greatly supersede human-level AGI? Here, a common misconception is that we know the answer with great certainty.

One popular myth is that we know we'll get superhuman AGI this century. In fact, history is full of technological over-hyping. Where are those fusion power plants and flying cars we were promised we'd have by now? AI too has been repeatedly over-hyped in the past, even by some of the founders of the field: for example, John McCarthy (who coined the term "artificial intelligence"), Marvin Minsky, Nathaniel Rochester and Claude Shannon wrote this overly optimistic forecast about what could be accomplished during two months with stone-age computers: "We propose that a 2 month, 10 man study of artificial intelligence be carried out during the summer of 1956 at Dartmouth College . . . An attempt will be made to find how to make machines use language, form abstractions and concepts, solve kinds of problems now reserved for humans, and improve themselves. We think that a significant advance can be made in one or more of these problems if a carefully selected group of scientists work on it together for a summer."

On the other hand, a popular counter-myth is that we know we *won't* get superhuman AGI this century. Researchers have made a wide range of estimates for how far we are from superhuman AGI, but we certainly can't say with great confidence that the probability is zero this century, given the dismal track record of such techno-skeptic predictions. For example, Ernest Rutherford, arguably the greatest nuclear physicist of his time, said in 1933—less than twenty-four hours before Leo Szilard's invention of the nuclear chain reaction—that nuclear energy was "moonshine," and in 1956 Astronomer Royal Richard Woolley called talk about space travel "utter bilge." The most extreme form of this myth is that superhuman AGI will never arrive because it's physically impossible. However, physicists know

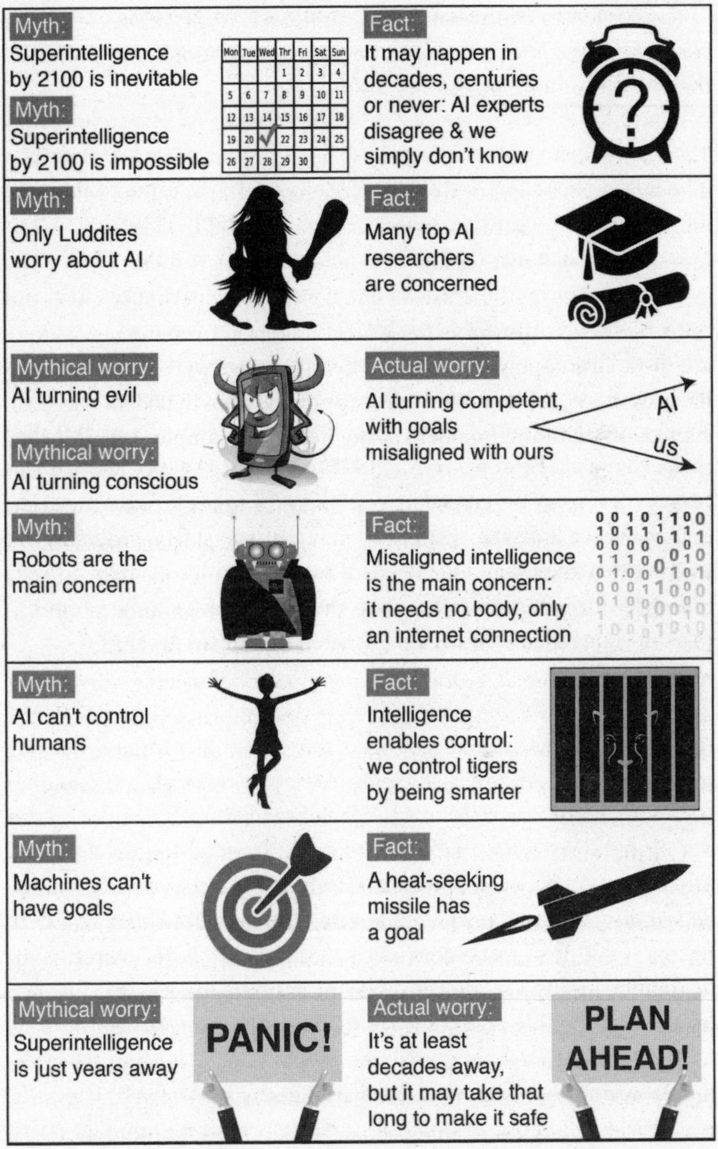

Myth: Superintelligence by 2100 is inevitable **Myth:** Superintelligence by 2100 is impossible	**Fact:** It may happen in decades, centuries or never: AI experts disagree & we simply don't know
Myth: Only Luddites worry about AI	**Fact:** Many top AI researchers are concerned
Mythical worry: AI turning evil **Mythical worry:** AI turning conscious	**Actual worry:** AI turning competent, with goals misaligned with ours (AI / us)
Myth: Robots are the main concern	**Fact:** Misaligned intelligence is the main concern: it needs no body, only an internet connection
Myth: AI can't control humans	**Fact:** Intelligence enables control: we control tigers by being smarter
Myth: Machines can't have goals	**Fact:** A heat-seeking missile has a goal
Mythical worry: Superintelligence is just years away — PANIC!	**Actual worry:** It's at least decades away, but it may take that long to make it safe — PLAN AHEAD!

Figure 1.5: Common myths about superintelligent AI.

that a brain consists of quarks and electrons arranged to act as a powerful computer, and that there's no law of physics preventing us from building even more intelligent quark blobs.

There have been a number of surveys asking AI researchers how many years from now they think we'll have human-level AGI with at least 50% probability, and all these surveys have the same conclusion: the world's leading experts disagree, so we simply don't know. For example, in such a poll of the AI researchers at the Puerto Rico AI conference, the average (median) answer was by the year 2055, but some researchers guessed hundreds of years or more.

There's also a related myth that people who worry about AI think it's only a few years away. In fact, most people on record worrying about superhuman AGI guess it's still at least decades away. But they argue that as long as we're not 100% *sure* that it won't happen this century, it's smart to start safety research now to prepare for the eventuality. As we'll see in this book, many of the safety problems are so hard that they may take decades to solve, so it's prudent to start researching them now rather than the night before some programmers drinking Red Bull decide to switch on human-level AGI.

Controversy Myths

Another common misconception is that the only people harboring concerns about AI and advocating AI-safety research are Luddites who don't know much about AI. When Stuart Russell mentioned this during his Puerto Rico talk, the audience laughed loudly. A related misconception is that supporting AI-safety research is hugely controversial. In fact, to support a modest investment in AI-safety research, people don't need to be convinced that risks are high, merely non-negligible, just as a modest investment in home insurance is justified by a non-negligible probability of the home burning down.

My personal analysis is that the media have made the AI-safety debate seem more controversial than it really is. After all, fear sells, and articles using out-of-context quotes to proclaim imminent doom can generate more clicks than nuanced and balanced ones. As a result, two people who only know about each other's positions from media quotes are likely to think they disagree more than they really do. For example, a techno-skeptic whose only knowledge about Bill Gates' position comes from a British tabloid may mistakenly think he believes superintelligence to be imminent. Similarly, someone in the beneficial-AI movement who knows nothing about Andrew Ng's position except his above-mentioned quote about overpopulation on Mars may mistakenly think he doesn't care about AI safety. In fact, I personally know that he does—the crux is simply that because his timeline estimates are longer, he naturally tends to prioritize short-term AI challenges over long-term ones.

Myths About What the Risks Are

I rolled my eyes when seeing this headline in the *Daily Mail*:[3] "Stephen Hawking Warns That Rise of Robots May Be Disastrous for Mankind." I've lost count of how many similar articles I've seen. Typically, they're accompanied by an evil-looking robot carrying a weapon, and suggest that we should worry about robots rising up and killing us because they've become conscious and/or evil. On a lighter note, such articles are actually rather impressive, because they succinctly summarize the scenario that my AI colleagues *don't* worry about. That scenario combines as many as three separate misconceptions: concern about *consciousness*, *evil* and *robots*, respectively.

If you drive down the road, you have a subjective experience of colors, sounds, etc. But does a self-driving car have a subjective experience? Does it feel like anything at all to be a self-driving car, or is it like an unconscious zombie without any subjective experience? Although this mystery of consciousness is interesting in its own right, and we'll devote chapter 8 to it, it's irrelevant to AI risk. If you get struck by a driverless car, it makes no difference to you whether it subjectively feels conscious. In the same way, what will affect us humans is what superintelligent AI *does*, not how it subjectively feels.

The fear of machines turning evil is another red herring. The real worry isn't malevolence, but competence. A superintelligent AI is by definition very good at attaining its goals, whatever they may be, so we need to ensure that its goals are aligned with ours. You're probably not an ant hater who steps on ants out of malice, but if you're in charge of a hydroelectric green energy project and there's an anthill in the region to be flooded, too bad for the ants. The beneficial-AI movement wants to avoid placing humanity in the position of those ants.

The consciousness misconception is related to the myth that machines can't have goals. Machines can obviously have goals in the narrow sense of exhibiting goal-oriented behavior: the behavior of a heat-seeking missile is most economically explained as a goal to hit a target. If you feel threatened by a machine whose goals are misaligned with yours, then it's precisely its goals in this narrow sense that trouble you, not whether the machine is conscious and experiences a sense of purpose. If that heat-seeking missile were chasing you, you probably wouldn't exclaim "I'm not worried, because machines can't have goals!"

I sympathize with Rodney Brooks and other robotics pioneers who feel unfairly demonized by scaremongering tabloids, because some journalists seem obsessively fixated on robots and adorn many of their articles with evil-looking metal monsters with shiny red eyes. In fact, the main concern of the beneficial-AI movement isn't with robots but with intelligence itself: specifically, intelligence whose goals are misaligned with ours. To cause us trouble, such misaligned intelligence needs no robotic body, merely an internet connection—we'll explore in chapter 4 how this may enable outsmarting financial markets, out-inventing human researchers, out-manipulating human leaders and developing weapons we cannot even understand. Even if building robots were physically impossible, a super-intelligent and super-wealthy AI could easily pay or manipulate myriad humans to unwittingly do its bidding, as in William Gibson's science fiction novel *Neuromancer*.

The robot misconception is related to the myth that machines can't control humans. Intelligence enables control: humans control tigers not because we're stronger, but because we're smarter. This means that if we cede our position as smartest on our planet, it's possible that we might also cede control.

Figure 1.5 summarizes all of these common misconceptions, so that we can dispense with them once and for all and focus our discussions with friends and colleagues on the many legitimate controversies—which, as we'll see, there's no shortage of!

The Road Ahead

In the rest of this book, you and I will explore together the future of life with AI. Let's navigate this rich and multifaceted topic in an organized way by first exploring the full story of life conceptually and chronologically, and then exploring goals, meaning and what actions to take to create the future we want.

In chapter 2, we explore the foundations of intelligence and how seemingly dumb matter can be rearranged to remember, compute and learn. As we proceed into the future, our story branches out into many scenarios defined by the answers to certain key questions. Figure 1.6 summarizes key questions we'll encounter as we march forward in time, to potentially ever more advanced AI.

Right now, we face the choice of whether to start an AI arms race, and questions about how to make tomorrow's AI systems bug-free and robust. If AI's economic impact keeps growing, we also have to decide how to modernize our laws and what career advice to give kids so that they can avoid soon-to-be-automated jobs. We explore such short-term questions in chapter 3.

If AI progress continues to human levels, then we also need to ask ourselves how to ensure that it's beneficial, and whether we can or should create a leisure society that flourishes without jobs. This also raises the question of whether an intelligence explosion or slow-but-steady growth can propel AGI far beyond human levels. We explore a wide range of such scenarios in chapter 4 and investigate the spectrum of possibilities for the aftermath in chapter 5, ranging from arguably dystopic to arguably utopic. Who's in charge—humans, AI or cyborgs? Are humans treated well or badly? Are we replaced and, if so, do we perceive our replacements as conquerors or worthy descendants? I'm very curious about which of the chapter 5 scenarios you personally prefer! I've set up a website, http://AgeOfAi.org, where you can share your views and join the conversation.

Finally, we forge billions of years into the future in chapter 6 where we can, ironically, draw stronger conclusions than in the previous chapters, as the ultimate limits of life in our cosmos are set not by intelligence but by the laws of physics.

After concluding our exploration of the history of intelligence,

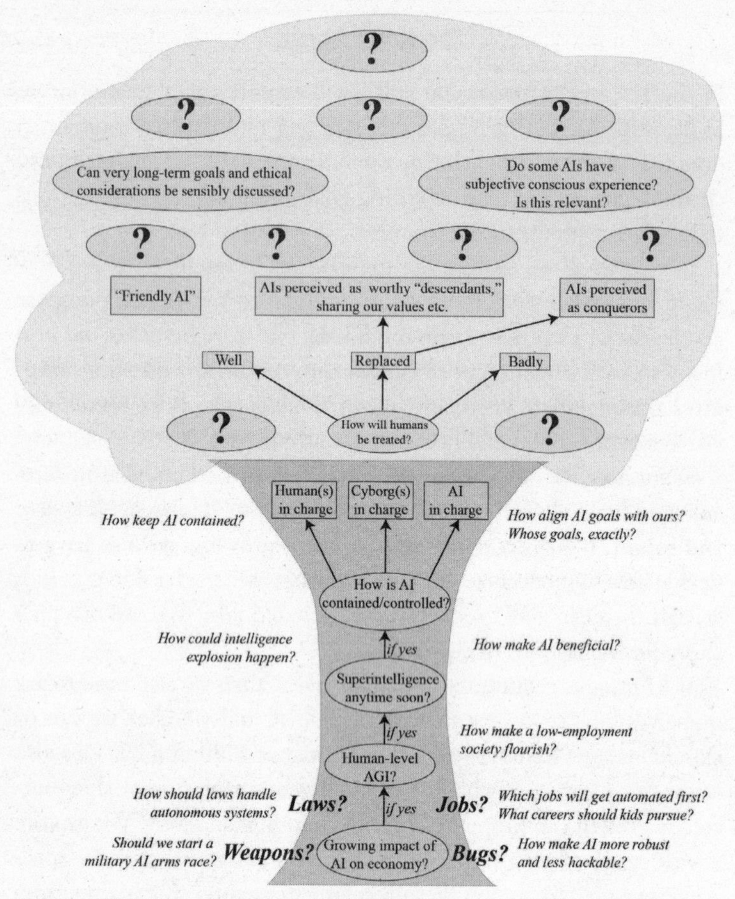

Figure 1.6: Which AI questions are interesting depends on how advanced AI gets and which branch our future takes.

we'll devote the remainder of the book to considering what future to aim for and how to get there. To be able to link cold facts to questions of purpose and meaning, we explore the physical basis of goals in chapter 7 and consciousness in chapter 8. Finally, in the epilogue, we explore what can be done right now to help create the future we want.

In case you're a reader who likes skipping around, most chapters are relatively self-contained once you've digested the terminology and definitions from this first chapter and the beginning of the next

		Short Chapter Title	Topic	Status
		Prelude: Tale of the Omega Team	Food for thought	Extremely Speculative
The history of intelligence	1	The Conversation	Key ideas, terminology	Not very speculative
	2	Matter Turns Intelligent	Fundamentals of intelligence	
	3	AI, Economics, Weapons & Law	Near future	
	4	Intelligence Explosion?	Superintelligence scenarios	Extremely Speculative
	5	Aftermath	Subsequent 10,000 years	
	6	Our Cosmic Endowment	Subsequent billions of years	
The history of meaning	7	Goals	History of goals-oriented behavior	Not very speculative
	8	Consciousness	Natural & artificial consciousness	Speculative
		Epilogue: Tale of the FLI Team	What should we do?	Not very speculative

Figure 1.7: Structure of the book

one. If you're an AI researcher, you can optionally skip all of chapter 2 except for its initial intelligence definitions. If you're new to AI, chapters 2 and 3 will give you the arguments for why chapters 4 through 6 can't be trivially dismissed as impossible science fiction. Figure 1.7 summarizes where the various chapters fall on the spectrum from factual to speculative.

A fascinating journey awaits us. Let's begin!

THE BOTTOM LINE:

- Life, defined as a process that can retain its complexity and replicate, can develop through three stages: a biological stage (1.0), where its hardware and software are evolved, a cultural stage (2.0), where it can design its software (through learning) and a technological stage (3.0), where it can design its hardware as well, becoming the master of its own destiny.
- Artificial intelligence may enable us to launch Life 3.0 this century, and a fascinating conversation has sprung up regarding what future we should aim for and how this can be accomplished. There are three main camps in the controversy: techno-skeptics, digital utopians and the beneficial-AI movement.
- Techno-skeptics view building superhuman AGI as so hard that it won't happen for hundreds of years, making it silly to worry about it (and Life 3.0) now.
- Digital utopians view it as likely this century and wholeheartedly welcome Life 3.0, viewing it as the natural and desirable next step in the cosmic evolution.
- The beneficial-AI movement also views it as likely this century, but views a good outcome not as guaranteed, but as something that needs to be ensured by hard work in the form of AI-safety research.
- Beyond such legitimate controversies where world-leading experts disagree, there are also boring pseudo-controversies caused by misunderstandings. For example, never waste time arguing about "life," "intelligence," or "consciousness" before ensuring that you and your protagonist are using these words to mean the same thing! This book uses the definitions in table 1.1.
- Also beware the common misconceptions in figure 1.5: "Superintelligence by 2100 is inevitable/impossible." "Only Luddites worry about AI." "The concern is about AI turning evil and/or conscious, and it's just years away." "Robots are the main concern." "AI can't control humans and can't have goals."
- In chapters 2 through 6, we'll explore the story of intelligence from its humble beginning billions of years ago to possible cosmic futures billions of years from now. We'll first investigate near-term challenges such as jobs, AI weapons and the quest for human-level AGI, then explore possibilities for a fascinating spectrum of possible futures with intelligent machines and/or humans. I wonder which options you'll prefer!
- In chapters 7 through 9, we'll switch from cold factual descriptions to an exploration of goals, consciousness and meaning, and investigate what we can do right now to help create the future we want.
- I view this conversation about the future of life with AI as the most important one of our time—please join it!

Chapter 2

Matter Turns Intelligent

Hydrogen . . . , given enough time, turns into people.

Edward Robert Harrison, 1995

One of the most spectacular developments during the 13.8 billion years since our Big Bang is that dumb and lifeless matter has turned intelligent. How could this happen and how much smarter can things get in the future? What does science have to say about the history and fate of intelligence in our cosmos? To help us tackle these questions, let's devote this chapter to exploring the foundations and fundamental building blocks of intelligence. What does it mean to say that a blob of matter is intelligent? What does it mean to say that an object can remember, compute and learn?

What Is Intelligence?

My wife and I recently had the good fortune to attend a symposium on artificial intelligence organized by the Swedish Nobel Foundation, and when a panel of leading AI researchers were asked to define intelligence, they argued at length without reaching consensus. We found this quite funny: there's no agreement on what intelligence is even among intelligent intelligence researchers! So there's clearly no undisputed "correct" definition of intelligence. Instead, there are many competing ones, including capacity for logic, understanding, planning, emotional knowledge, self-awareness, creativity, problem solving and learning.

In our exploration of the future of intelligence, we want to take a maximally broad and inclusive view, not limited to the sorts of intelligence that exist so far. That's why the definition I gave in the last chapter, and the way I'm going to use the word throughout this book, is very broad:

intelligence = *ability to accomplish complex goals*

This is broad enough to include all above-mentioned definitions, since understanding, self-awareness, problem solving, learning, etc. are all examples of complex goals that one might have. It's also broad enough to subsume the *Oxford Dictionary* definition—"the ability to acquire and apply knowledge and skills"—since one can have as a goal to apply knowledge and skills.

Because there are many possible goals, there are many possible types of intelligence. By our definition, it therefore makes no sense to quantify intelligence of humans, non-human animals or machines by a single number such as an IQ.* What's more intelligent: a computer program that can only play chess or one that can only play Go? There's no sensible answer to this, since they're good at different things that can't be directly compared. We can, however, say that a third program is more intelligent than both of the others if it's at least as good as them at accomplishing *all* goals, and strictly better at at least one (winning at chess, say).

It also makes little sense to quibble about whether something is or isn't intelligent in borderline cases, since ability comes on a spectrum and isn't necessarily an all-or-nothing trait. What people have the ability to accomplish the goal of speaking? Newborns? No. Radio hosts? Yes. But what about toddlers who can speak ten words? Or five hundred words? Where would you draw the line? I've used the deliberately vague word "complex" in the definition above, because it's not very interesting to try to draw an artificial line between intel-

* To see this, imagine how you'd react if someone claimed that the ability to accomplish Olympic-level athletic feats could be quantified by a single number called the "athletic quotient," or AQ for short, so that the Olympian with the highest AQ would win the gold medals in all the sports.

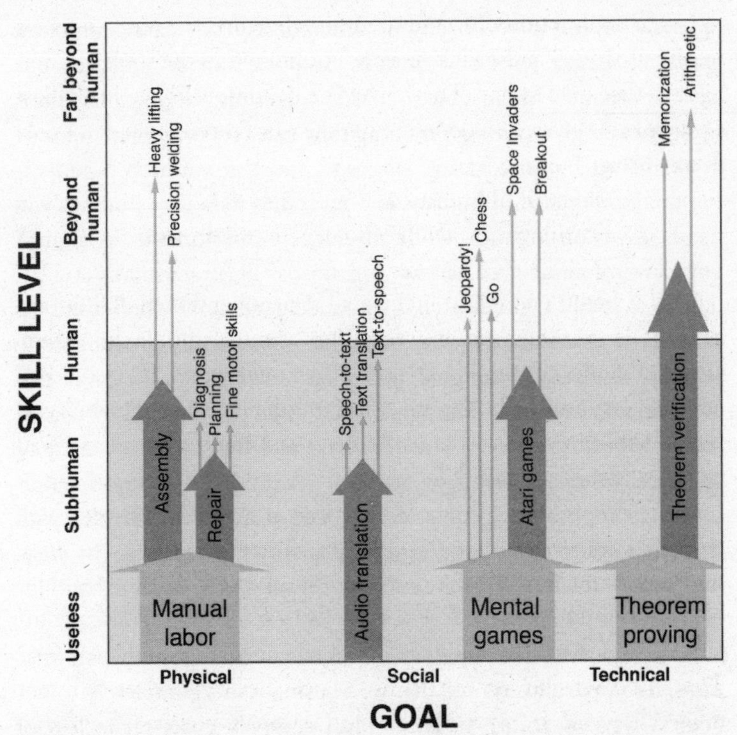

Figure 2.1: Intelligence, defined as ability to accomplish complex goals, can't be measured by a single IQ, only by an ability spectrum across all goals. Each arrow indicates how skilled today's best AI systems are at accomplishing various goals, illustrating that today's artificial intelligence tends to be *narrow*, with each system able to accomplish only very specific goals. In contrast, human intelligence is remarkably broad: a healthy child can learn to get better at almost anything.

ligence and non-intelligence, and it's more useful to simply quantify the degree of ability for accomplishing different goals.

To classify different intelligences into a taxonomy, another crucial distinction is that between *narrow* and *broad* intelligence. IBM's Deep Blue chess computer, which dethroned chess champion Garry Kasparov in 1997, was only able to accomplish the very narrow task of playing chess—despite its impressive hardware and software, it couldn't even beat a four-year-old at tic-tac-toe. The DQN AI system of Google DeepMind can accomplish a slightly broader range of goals: it can play dozens of different vintage Atari computer games at human level or better. In contrast, human intelligence is thus far uniquely broad, able to master a dazzling panoply of skills. A healthy child given enough training time can get fairly good not only at *any* game, but also at any language, sport or vocation. Comparing the intelligence of humans and machines today, we humans win hands-down on breadth, while machines outperform us in a small but growing number of narrow domains, as illustrated in figure 2.1. The holy grail of AI research is to build "general AI" (better known as *artificial general intelligence*, AGI) that is maximally broad: able to accomplish virtually any goal, including learning. We'll explore this in detail in chapter 4. The term "AGI" was popularized by the AI researchers Shane Legg, Mark Gubrud and Ben Goertzel to more specifically mean *human-level* artificial general intelligence: the ability to accomplish any goal at least as well as humans.[1] I'll stick with their definition, so unless I explicitly qualify the acronym (by writing "superhuman AGI," for example), I'll use "AGI" as shorthand for "human-level AGI."*

Although the word "intelligence" tends to have positive connotations, it's important to note that we're using it in a completely value-neutral way: as ability to accomplish complex goals regardless of whether these goals are considered good or bad. Thus an intelligent person may be very good at helping people or very good at hurting people. We'll explore the issue of goals in chapter 7. Regarding goals, we also need to clear up the subtlety of whose goals we're referring to. Suppose your future brand-new robotic personal assistant has no goals whatsoever of its own, but will do whatever you ask it to do, and you ask it to cook the perfect Italian dinner. If it goes online and researches Italian dinner recipes, how to get to the closest supermarket, how to strain pasta and so on, and then successfully buys the ingredients and prepares a succulent meal, you'll presumably consider it intelligent even though the original goal was yours. In fact, it adopted your goal once you'd made your request, and then broke

* Some people prefer "human-level AI" or "strong AI" as synonyms for AGI, but both are problematic. Even a pocket calculator is a human-level AI in the narrow sense. The antonym of "strong AI" is "weak AI," but it feels odd to call narrow AI systems such as Deep Blue, Watson, and AlphaGo "weak."

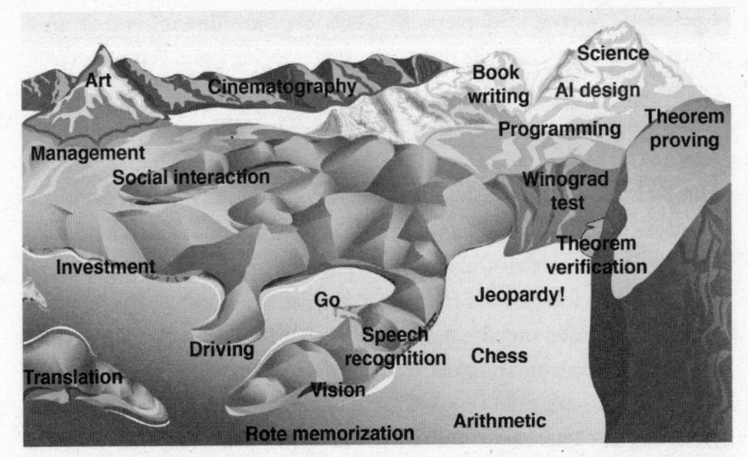

Figure 2.2: Illustration of Hans Moravec's "landscape of human competence," where elevation represents difficulty for computers, and the rising sea level represents what computers are able to do.

it into a hierarchy of subgoals of its own, from paying the cashier to grating the Parmesan. In this sense, intelligent behavior is inexorably linked to goal attainment.

It's natural for us to rate the difficulty of tasks relative to how hard it is for us humans to perform them, as in figure 2.1. But this can give a misleading picture of how hard they are for computers. It feels much harder to multiply 314,159 by 271,828 than to recognize a friend in a photo, yet computers creamed us at arithmetic long before I was born, while human-level image recognition has only recently become possible. This fact that low-level sensorimotor tasks seem easy despite requiring enormous computational resources is known as Moravec's paradox, and is explained by the fact that our brain makes such tasks feel easy by dedicating massive amounts of customized hardware to them—more than a quarter of our brains, in fact.

I love this metaphor from Hans Moravec, and have taken the liberty to illustrate it in figure 2.2:

> Computers are universal machines, their potential extends uniformly over a boundless expanse of tasks. Human potentials, on the other hand, are strong in areas long important for survival, but weak in things far removed. Imagine a "landscape of human com-

> petence," having lowlands with labels like "arithmetic" and "rote memorization," foothills like "theorem proving" and "chess playing," and high mountain peaks labeled "locomotion," "hand-eye coordination" and "social interaction." Advancing computer performance is like water slowly flooding the landscape. A half century ago it began to drown the lowlands, driving out human calculators and record clerks, but leaving most of us dry. Now the flood has reached the foothills, and our outposts there are contemplating retreat. We feel safe on our peaks, but, at the present rate, those too will be submerged within another half century. I propose that we build Arks as that day nears, and adopt a seafaring life![2]

During the decades since he wrote those passages, the sea level has kept rising relentlessly, as he predicted, like global warming on steroids, and some of his foothills (including chess) have long since been submerged. What comes next and what we should do about it is the topic of the rest of this book.

As the sea level keeps rising, it may one day reach a tipping point, triggering dramatic change. This critical sea level is the one corresponding to machines becoming able to perform AI design. Before this tipping point is reached, the sea-level rise is caused by *humans* improving machines; afterward, the rise can be driven by *machines* improving machines, potentially much faster than humans could have done, rapidly submerging all land. This is the fascinating and controversial idea of the *singularity*, which we'll have fun exploring in chapter 4.

Computer pioneer Alan Turing famously proved that if a computer can perform a certain bare minimum set of operations, then, given enough time and memory, it can be programmed to do anything that *any* other computer can do. Machines exceeding this critical threshold are called *universal computers* (aka Turing-universal computers); all of today's smartphones and laptops are universal in this sense. Analogously, I like to think of the critical intelligence threshold required for AI design as the threshold for *universal intelligence:* given enough time and resources, it can make itself able to accomplish any goal as well as *any* other intelligent entity. For example, if it decides that it wants better social skills, forecasting skills or AI-design skills, it can acquire them. If it decides to figure out how to build a robot factory, then it can do so. In other words, universal intelligence has the potential to develop into Life 3.0.

The conventional wisdom among artificial intelligence researchers is that intelligence is ultimately all about information and computation, not about flesh, blood or carbon atoms. This means that there's no fundamental reason why machines can't one day be at least as intelligent as us.

But what are information and computation really, given that physics has taught us that, at a fundamental level, everything is simply matter and energy moving around? How can something as abstract, intangible and ethereal as information and computation be embodied by tangible physical stuff? In particular, how can a bunch of dumb particles moving around according to the laws of physics exhibit behavior that we'd call intelligent?

If you feel that the answer to this question is obvious and consider it plausible that machines might get as intelligent as humans this century—for example because you're an AI researcher—please skip the rest of this chapter and jump straight to chapter 3. Otherwise, you'll be pleased to know that I've written the next three sections specially for you.

What Is Memory?

If we say that an atlas contains *information* about the world, we mean that there's a relation between the state of the book (in particular, the positions of certain molecules that give the letters and images their colors) and the state of the world (for example, the locations of continents). If the continents were in different places, then those molecules would be in different places as well. We humans use a panoply of different devices for storing information, from books and brains to hard drives, and they all share this property: that their state can be related to (and therefore inform us about) the state of other things that we care about.

What fundamental physical property do they all have in common that makes them useful as memory devices, i.e., devices for storing information? The answer is that they all *can be in many different long-*

lived states—long-lived enough to encode the information until it's needed. As a simple example, suppose you place a ball on a hilly surface that has sixteen different valleys, as in figure 2.3. Once the ball has rolled down and come to rest, it will be in one of sixteen places, so you can use its position as a way of remembering any number between 1 and 16.

This memory device is rather robust, because even if it gets a bit jiggled and disturbed by outside forces, the ball is likely to stay in the same valley that you put it in, so you can still tell which number is being stored. The reason that this memory is so stable is that lifting the ball out of its valley requires more energy than random disturbances are likely to provide. This same idea can provide stable memories much more generally than for a movable ball: the energy of a complicated physical system can depend on all sorts of mechanical, chemical, electrical and magnetic properties, and as long as it takes energy to change the system away from the state you want it to remember, this state will be stable. This is why solids have many long-lived states, whereas liquids and gases don't: if you engrave someone's name on a gold ring, the information will still be there years later because reshaping the gold requires significant energy, but if you engrave it in the surface of a pond, it will be lost within a second as the water surface effortlessly changes its shape.

The simplest possible memory device has only two stable states (figure 2.3, page 57). We can therefore think of it as encoding a binary digit (abbreviated "bit"), i.e., a zero or a one. The information stored by any more complicated memory device can equivalently be stored in multiple bits: for example, taken together, the four bits shown in figure 2.3 (right) can be in $2 \times 2 \times 2 \times 2 = 16$ different states 0000, 0001, 0010, 0011, . . . , 1111, so they collectively have exactly the same memory capacity as the more complicated 16-state system (left). We can therefore think of bits as atoms of information—the smallest indivisible chunk of information that can't be further subdivided, which can combine to make up any information. For example, I just typed the word "word," and my laptop represented it in its memory as the 4-number sequence 119 111 114 100, storing each of those numbers as 8 bits (it represents each lowercase letter by a number that's 96 plus its order in the alphabet). As soon as I hit the *w* key on

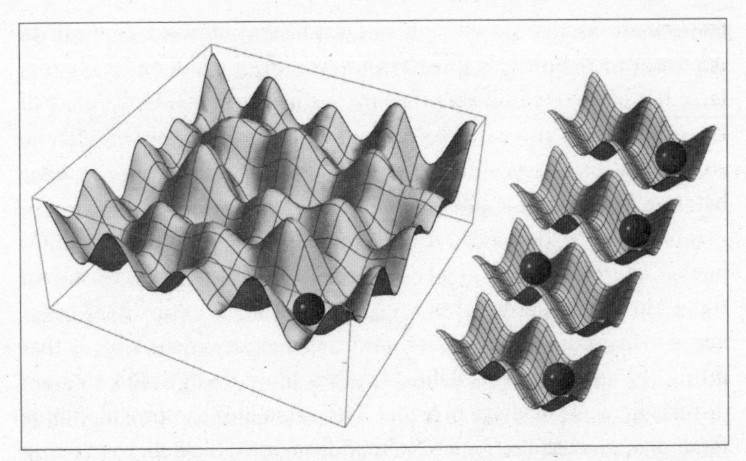

Figure 2.3: A physical object is a useful memory device if it can be in many different stable states. The ball on the left can encode four bits of information labeling which one of $2^4 = 16$ valleys it's in. Together, the four balls on the right also encode four bits of information—one bit each.

my keyboard, my laptop displayed a visual image of a *w* on my screen, and this image is also represented by bits: 32 bits specify the color of each of the screen's millions of pixels.

Since two-state systems are easy to manufacture and work with, most modern computers store their information as bits, but these bits are embodied in a wide variety of ways. On a DVD, each bit corresponds to whether there is or isn't a microscopic pit at a given point on the plastic surface. On a hard drive, each bit corresponds to a point on the surface being magnetized in one of two ways. In my laptop's working memory, each bit corresponds to the positions of certain electrons, determining whether a device called a micro-capacitor is charged. Some kinds of bits are convenient to transport as well, even at the speed of light: for example, in an optical fiber transmitting your email, each bit corresponds to a laser beam being strong or weak at a given time.

Engineers prefer to encode bits into systems that aren't only stable and easy to read from (as a gold ring), but also easy to write to: altering the state of your hard drive requires much less energy than engraving gold. They also prefer systems that are convenient to work with and cheap to mass-produce. But other than that, they simply don't care about how the bits are represented as physical objects—and nor do you most of the time, because it simply doesn't matter! If you email your friend a document to print, the information may get copied in rapid succession from magnetizations on your hard drive to electric charges in your computer's working memory, radio waves in your wireless network, voltages in your router, laser pulses in an optical fiber and, finally, molecules on a piece of paper. In other words, *information can take on a life of its own, independent of its physical substrate!* Indeed, it's usually only this substrate-independent aspect of information that we're interested in: if your friend calls you up to discuss that document you sent, she's probably not calling to talk about voltages or molecules. This is our first hint of how something as intangible as intelligence can be embodied in tangible physical stuff, and we'll soon see how this idea of substrate independence is much deeper, including not only information but also computation and learning.

Because of this substrate independence, clever engineers have been able to repeatedly replace the memory devices inside our computers with dramatically better ones, based on new technologies, without requiring any changes whatsoever to our software. The result has been spectacular, as illustrated in figure 2.4: over the past six decades, computer memory has gotten half as expensive roughly every couple of years. Hard drives have gotten over 100 million times cheaper, and the faster memories useful for computation rather than mere storage have become a whopping 10 trillion times cheaper. If you could get such a "99.99999999999% off" discount on all your shopping, you could buy all real estate in New York City for about 10 cents and all the gold that's ever been mined for around a dollar.

For many of us, the spectacular improvements in memory technology come with personal stories. I fondly remember working in a candy store back in high school to pay for a computer sporting 16 kilobytes of memory, and when I made and sold a word processor for it with my high school classmate Magnus Bodin, we were forced to write it all in ultra-compact machine code to leave enough memory for the words that it was supposed to process. After getting used to floppy drives storing 70kB, I became awestruck by the smaller 3.5-inch floppies that could store a whopping 1.44MB and hold a whole

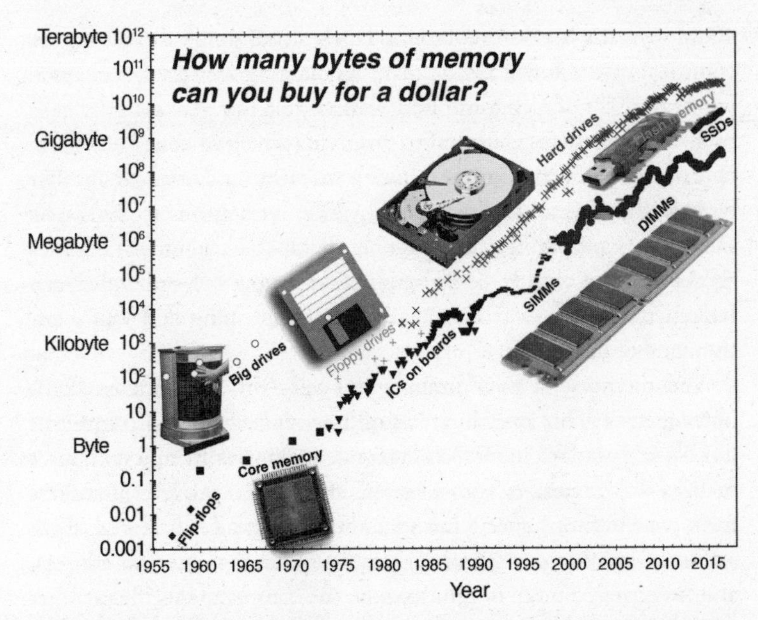

Figure 2.4: Over the past six decades, computer memory has gotten twice as cheap roughly every couple of years, corresponding to a thousand times cheaper roughly every twenty years. A byte equals eight bits. Data courtesy of John McCallum, from http://www.jcmit.net/memoryprice.htm.

book, and then my first-ever hard drive storing 10MB—which might just barely fit a single one of today's song downloads. These memories from my adolescence felt almost unreal the other day, when I spent about $100 on a hard drive with 300,000 times more capacity.

What about memory devices that evolved rather than being designed by humans? Biologists don't yet know what the first-ever life form was that copied its blueprints between generations, but it may have been quite small. A team led by Philipp Holliger at Cambridge University made an RNA molecule in 2016 that encoded 412 bits of genetic information and was able to copy RNA strands longer than itself, bolstering the "RNA world" hypothesis that early Earth life involved short self-replicating RNA snippets. So far, the smallest memory device known to be evolved and used in the wild is the genome of the bacterium Candidatus Carsonella ruddii, storing about

40 kilobytes, whereas our human DNA stores about 1.6 gigabytes, comparable to a downloaded movie. As mentioned in the last chapter, our brains store much more information than our genes: in the ballpark of 10 gigabytes electrically (specifying which of your 100 billion neurons are firing at any one time) and 100 terabytes chemically/biologically (specifying how strongly different neurons are linked by synapses). Comparing these numbers with the machine memories shows that the world's best computers can now out-remember any biological system—at a cost that's rapidly dropping and was a few thousand dollars in 2016.

The memory in your brain works very differently from computer memory, not only in terms of how it's built, but also in terms of how it's used. Whereas you retrieve memories from a computer or hard drive by specifying *where* it's stored, you retrieve memories from your brain by specifying something about *what* is stored. Each group of bits in your computer's memory has a numerical address, and to retrieve a piece of information, the computer specifies at what address to look, just as if I tell you "Go to my bookshelf, take the fifth book from the right on the top shelf, and tell me what it says on page 314." In contrast, you retrieve information from your brain similarly to how you retrieve it from a search engine: you specify a piece of the information or something related to it, and it pops up. If I tell you "to be or not," or if I google it, chances are that it will trigger "To be, or not to be, that is the question." Indeed, it will probably work even if I use another part of the quote or mess things up somewhat. Such memory systems are called *auto-associative*, since they recall by association rather than by address.

In a famous 1982 paper, the physicist John Hopfield showed how a network of interconnected neurons could function as an auto-associative memory. I find the basic idea very beautiful, and it works for any physical system with multiple stable states. For example, consider a ball on a surface with two valleys, like the one-bit system in figure 2.3, and let's shape the surface so that the x-coordinates of the two minima where the ball can come to rest are $x = \sqrt{2} \approx 1.41421$ and $x = \pi \approx 3.14159$, respectively. If you remember only that π is close to 3, you simply put the ball at $x = 3$ and watch it reveal a more exact π-value as it rolls down to the nearest minimum. Hopfield realized that a complex network of neurons provides an analogous landscape with very many energy-minima that the system can settle into, and it was later proved that you can squeeze in as many as 138 different memories for every thousand neurons without causing major confusion.

What Is Computation?

We've now seen how a physical object can remember information. But how can it compute?

A computation is a transformation of one memory state into another. In other words, a computation takes information and transforms it, implementing what mathematicians call a *function*. I think of a function as a meat grinder for information, as illustrated in figure 2.5: you put information in at the top, turn the crank and get processed information out at the bottom—and you can repeat this as many times as you want with different inputs. This information processing is deterministic in the sense that if you repeat it with the same input, you get the same output every time.

Although it sounds deceptively simple, this idea of a function is incredibly general. Some functions are rather trivial, such as the one

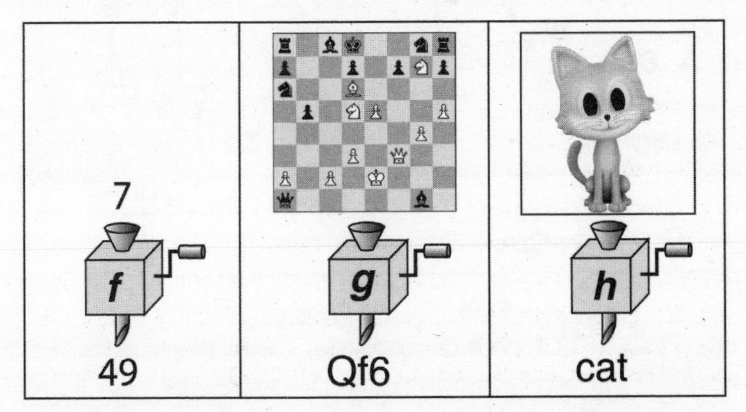

Figure 2.5: A *computation* takes information and transforms it, implementing what mathematicians call a *function*. The function f (left) takes bits representing a number and computes its square. The function g (middle) takes bits representing a chess position and computes the best move for White. The function h (right) takes bits representing an image and computes a text label describing it.

called *NOT* that inputs a single bit and outputs the reverse, thus turning zero into one and vice versa. The functions we learn about in school typically correspond to buttons on a pocket calculator, inputting one or more numbers and outputting a single number—for example, the function x^2 simply inputs a number and outputs it multiplied by itself. Other functions can be extremely complicated. For instance, if you're in possession of a function that would input bits representing an arbitrary chess position and output bits representing the best possible next move, you can use it to win the World Computer Chess Championship. If you're in possession of a function that inputs all the world's financial data and outputs the best stocks to buy, you'll soon be extremely rich. Many AI researchers dedicate their careers to figuring out how to implement certain functions. For example, the goal of machine-translation research is to implement a function inputting bits representing text in one language and outputting bits representing that same text in another language, and the goal of automatic-captioning research is inputting bits representing an image and outputting bits representing text describing it (figure 2.5, right).

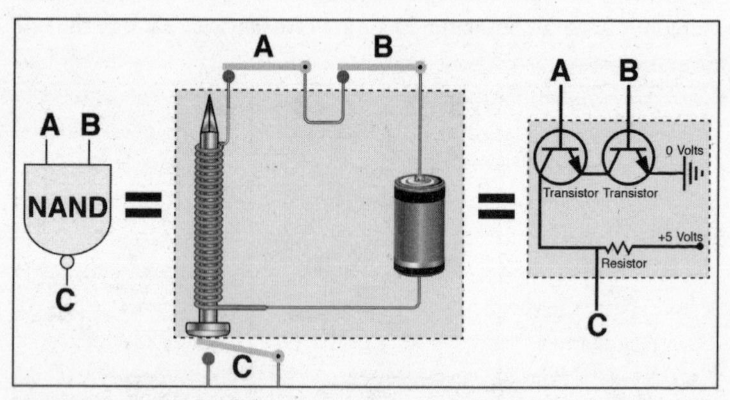

Figure 2.6: A so-called NAND gate takes two bits, A and B, as inputs and computes one bit C as output, according to the rule that C = 0 if A = B = 1 and C = 1 otherwise. Many physical systems can be used as NAND gates. In the middle example, switches are interpreted as bits where 0 = open, 1= closed, and when switches A and B are both closed, an electromagnet opens the switch C. In the rightmost example, voltages (electrical potentials) are interpreted as bits where 1 = five volts, 0 = zero volts, and when wires A and B are both at five volts, the two transistors conduct electricity and the wire C drops to approximately zero volts.

In other words, if you can implement highly complex functions, then you can build an intelligent machine that's able to accomplish highly complex goals. This brings our question of how matter can be intelligent into sharper focus: in particular, how can a clump of seemingly dumb matter compute a complicated function?

Rather than just remain immobile as a gold ring or other static memory device, it must exhibit complex *dynamics* so that its future state depends in some complicated (and hopefully controllable/programmable) way on the present state. Its atom arrangement must be less ordered than a rigid solid where nothing interesting changes, but more ordered than a liquid or gas. Specifically, we want the system to have the property that if we put it in a state that encodes the input information, let it evolve according to the laws of physics for some amount of time, and then interpret the resulting final state as the output information, then the output is the desired function of the input. If this is the case, then we can say that our system computes our function.

As a first example of this idea, let's explore how we can build a very simple (but also very important) function called a *NAND gate** out of plain old dumb matter. This function inputs two bits and outputs one bit: it outputs 0 if both inputs are 1; in all other cases, it outputs 1. If we connect two switches in series with a battery and an electromagnet, then the electromagnet will only be on if the first switch *and* the second switch are closed ("on"). Let's place a third switch under the electromagnet, as illustrated in figure 2.6, such that the magnet will pull it open whenever it's powered on. If we interpret the first two switches as the input bits and the third one as the output bit (with 0 = switch open, and 1 = switch closed), then we have ourselves a NAND gate: the third switch is open only if the first two are closed. There are many other ways of building NAND gates that are more practical—for example, using transistors as illustrated in figure 2.6 (page 62). In today's computers, NAND gates are typically built from microscopic transistors and other components that can be automatically etched onto silicon wafers.

* NAND is short for NOT AND: An AND gate outputs 1 only if the first input is 1 and the second input is 1, so NAND outputs the exact opposite.

There's a remarkable theorem in computer science that says that NAND gates are *universal*, meaning that you can implement *any* well-defined function simply by connecting together NAND gates.* So if you can build enough NAND gates, you can build a device computing anything! In case you'd like a taste of how this works, I've illustrated in figure 2.7 how to multiply numbers using nothing but NAND gates.

MIT researchers Norman Margolus and Tommaso Toffoli coined the name *computronium* for any substance that can perform arbitrary computations. We've just seen that making computronium doesn't have to be particularly hard: the substance just needs to be able to implement NAND gates connected together in any desired way. Indeed, there are myriad other kinds of computronium as well. A simple variant that also works involves replacing the NAND gates by NOR gates that output 1 only when both inputs are 0. In the next section, we'll explore neural networks, which can also implement arbitrary computations, i.e., act as computronium. Scientist and entrepreneur Stephen Wolfram has shown that the same goes for simple devices called cellular automata, which repeatedly update bits based on what neighboring bits are doing. Already back in 1936, computer pioneer Alan Turing proved in a landmark paper that a simple machine (now known as a "universal Turing machine") that could manipulate symbols on a strip of tape could also implement arbitrary computations. In summary, not only is it possible for matter to implement any well-defined computation, but it's possible in a plethora of different ways.

As mentioned earlier, Turing also proved something even more profound in that 1936 paper of his: that if a type of computer can perform a certain bare minimum set of operations, then it's *universal* in the sense that given enough resources, it can do anything that any other computer can do. He showed that his Turing machine was

* I'm using "well-defined function" to mean what mathematicians and computer scientists call a "computable function," i.e., a function that could be computed by some hypothetical computer with unlimited memory and time. Alan Turing and Alonzo Church famously proved that there are also functions that can be described but aren't computable.

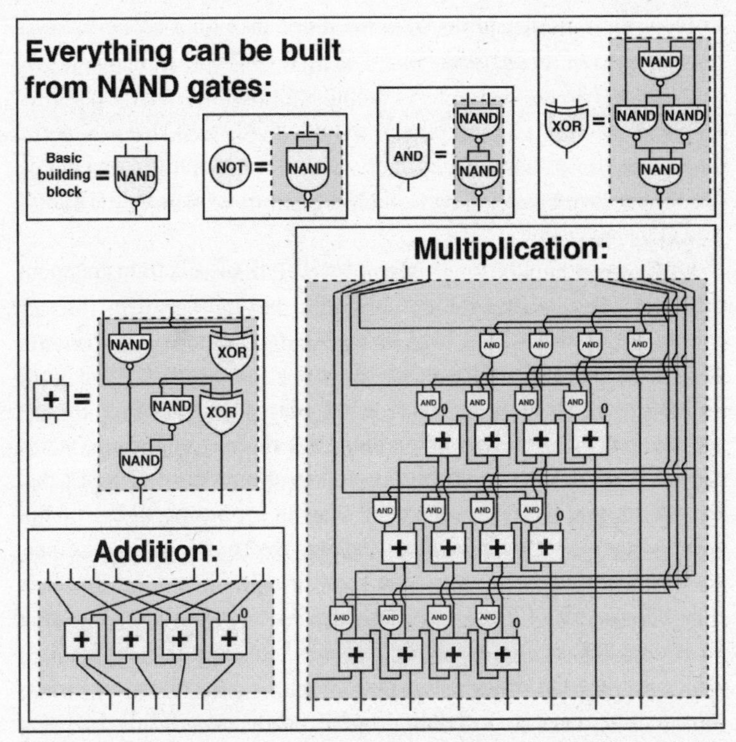

Figure 2.7: *Any* well-defined computation can be performed by cleverly combining nothing but NAND gates. For example, the addition and multiplication modules above both input two binary numbers represented by 4 bits, and output a binary number represented by 5 bits and 8 bits, respectively. The smaller modules NOT, AND, XOR and + (which sums three separate bits into a 2-bit binary number) are in turn built out of NAND gates. Fully understanding this figure is extremely challenging and totally unnecessary for following the rest of this book; I'm including it here just to illustrate the idea of universality—and to satisfy my inner geek.

universal, and connecting back more closely to physics, we've just seen that this family of universal computers also includes objects as diverse as a network of NAND gates and a network of interconnected neurons. Indeed, Stephen Wolfram has argued that *most* non-trivial physical systems, from weather systems to brains, would be universal computers if they could be made arbitrarily large and long-lasting.

This fact that exactly the same computation can be performed on *any* universal computer means that *computation is substrate-independent* in the same way that information is: it can take on a life of its own, independent of its physical substrate! So if you're a conscious superintelligent character in a future computer game, you'd have no way of knowing whether you ran on a Windows desktop, a Mac OS laptop or an Android phone, because you would be substrate-independent. You'd also have no way of knowing what type of transistors the microprocessor was using.

I first came to appreciate this crucial idea of substrate independence because there are many beautiful examples of it in physics. Waves, for instance: they have properties such as speed, wavelength and frequency, and we physicists can study the equations they obey without even needing to know what particular substance they're waves in. When you hear something, you're detecting sound waves caused by molecules bouncing around in the mixture of gases that we call air, and we can calculate all sorts of interesting things about these waves—how their intensity fades as the square of the distance, such as how they bend when they pass through open doors and how they bounce off of walls and cause echoes—without knowing what air is made of. In fact, we don't even need to know that it's made of molecules: we can ignore all details about oxygen, nitrogen, carbon dioxide, etc., because the only property of the wave's substrate that matters and enters into the famous wave equation is a single number that we can measure: the wave speed, which in this case is about 300 meters per second. Indeed, this wave equation that I taught my MIT students about in a course last spring was first discovered and put to great use long before physicists had even established that atoms and molecules existed!

This wave example illustrates three important points. First, substrate independence doesn't mean that a substrate is unnecessary, but that most of its details don't matter. You obviously can't have sound waves in a gas if there's no gas, but any gas whatsoever will suffice. Similarly, you obviously can't have computation without matter, but any matter will do as long as it can be arranged into NAND gates, connected neurons or some other building block enabling universal computation. Second, the substrate-independent phenomenon takes on a life of its own, independent of its substrate. A wave can travel across a lake, even though none of its water molecules do—they mostly bob up and down, like fans doing "the wave" in a sports stadium. Third, it's often only the substrate-independent aspect that we're interested in: a surfer usually cares more about the position and height of a wave than about its detailed molecular composition. We saw how this was true for information, and it's true for computation too: if two programmers are jointly hunting a bug in their code, they're probably not discussing transistors.

We've now arrived at an answer to our opening question about how tangible physical stuff can give rise to something that feels as intangible, abstract and ethereal as intelligence: it feels so non-physical because it's substrate-independent, taking on a life of its own that doesn't depend on or reflect the physical details. In short, computation is a pattern in the spacetime arrangement of particles, and it's not the particles but the pattern that really matters! Matter doesn't matter.

In other words, the hardware is the matter and the software is the pattern. This substrate independence of computation implies that AI is possible: intelligence doesn't require flesh, blood or carbon atoms.

Because of this substrate independence, shrewd engineers have been able to repeatedly replace the technologies inside our computers with dramatically better ones, without changing the software. The results have been every bit as spectacular as those for memory devices. As illustrated in figure 2.8, computation keeps getting half as expensive roughly every couple of years, and this trend has now persisted for over a century, cutting the computer cost a whopping million million million (10^{18}) times since my grandmothers were born. If everything got a million million million times cheaper, then a hundredth of a cent would enable you to buy all goods and services produced on Earth this year. This dramatic drop in costs is of course a key reason why computation is everywhere these days, having spread from the building-sized computing facilities of yesteryear into our homes, cars and pockets—and even turning up in unexpected places such as sneakers.

Why does our technology keep doubling its power at regular intervals, displaying what mathematicians call exponential growth?

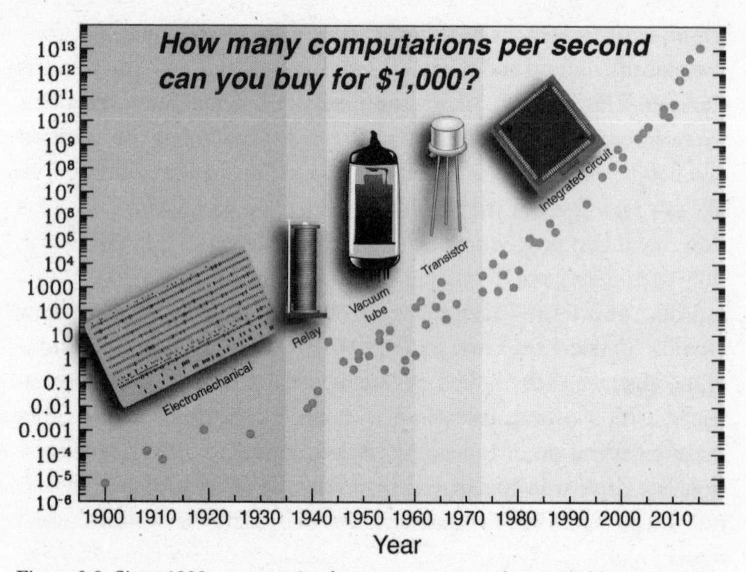

Figure 2.8: Since 1900, computation has gotten twice as cheap roughly every couple of years. The plot shows the computing power measured in floating-point operations per second (FLOPS) that can be purchased for $1,000.[3] The particular computation that defines a floating point operation corresponds to about 10^5 elementary logical operations such as bit flips or NAND evaluations.

Indeed, why is it happening not only in terms of transistor miniaturization (a trend known as *Moore's law*), but also more broadly for computation as a whole (figure 2.8), for memory (figure 2.4) and for a plethora of other technologies ranging from genome sequencing to brain imaging? Ray Kurzweil calls this persistent doubling phenomenon "the law of accelerating returns."

All examples of persistent doubling that I know of in nature have the same fundamental cause, and this technological one is no exception: each step creates the next. For example, you yourself underwent exponential growth right after your conception: each of your cells divided and gave rise to two cells roughly daily, causing your total number of cells to increase day by day as 1, 2, 4, 8, 16 and so on. According to the most popular scientific theory of our cosmic origins, known as *inflation*, our baby Universe once grew exponentially just like you did, repeatedly doubling its size at regular intervals until a speck much smaller and lighter than an atom had grown more massive than all the galaxies we've ever seen with our telescopes. Again, the cause was a process whereby each doubling step caused the next. This is how technology progresses as well: once technology gets twice as powerful, it can often be used to design and build technology that's twice as powerful in turn, triggering repeated capability doubling in the spirit of Moore's law.

Something that occurs just as regularly as the doubling of our technological power is the appearance of claims that the doubling is ending. Yes, Moore's law will of course end, meaning that there's a physical limit to how small transistors can be made. But some people mistakenly assume that Moore's law is synonymous with the persistent doubling of our technological power. Contrariwise, Ray Kurzweil points out that Moore's law involves not the first but the fifth technological paradigm to bring exponential growth in computing, as illustrated in figure 2.8: whenever one technology stopped improving, we replaced it with an even better one. When we could no longer keep shrinking our vacuum tubes, we replaced them with transistors and then integrated circuits, where electrons move around in two dimensions. When this technology reaches its limits, there are many other alternatives we can try—for example, using three-dimensional circuits and using something other than electrons to do our bidding.

Nobody knows for sure what the next blockbuster computational substrate will be, but we do know that we're nowhere near the limits imposed by the laws of physics. My MIT colleague Seth Lloyd has worked out what this fundamental limit is, and as we'll explore in greater detail in chapter 6, this limit is a whopping 33 orders of magnitude (10^{33} times) beyond today's state of the art for how much computing a clump of matter can do. So even if we keep doubling the power of our computers every couple of years, it will take over two centuries until we reach that final frontier.

Although all universal computers are capable of the same computations, some are more efficient than others. For example, a computation requiring millions of multiplications doesn't require millions of separate multiplication modules built from separate transistors as in figure 2.6: it needs only one such module, since it can use it many times in succession with appropriate inputs. In this spirit of efficiency, most modern computers use a paradigm where computations are split into multiple time steps, during which information is shuffled back and forth between memory modules and computation modules. This computational architecture was developed between 1935 and 1945 by computer pioneers including Alan Turing, Konrad Zuse, Presper Eckert, John Mauchly and John von Neumann. More specifically, the computer memory stores both data and software (a program, i.e., a list of instructions for what to do with the data). At each time step, a central processing unit (CPU) executes the next instruction in the program, which specifies some simple function to apply to some part of the data. The part of the computer that keeps track of what to do next is merely another part of its memory, called the *program counter*, which stores the current line number in the program. To go to the next instruction, simply add one to the program counter. To jump to another line of the program, simply copy that line number into the program counter—this is how so-called "if" statements and loops are implemented.

Today's computers often gain additional speed by *parallel processing*, which cleverly undoes some of this reuse of modules: if a computation can be split into parts that can be done in parallel (because the input of one part doesn't require the output of another), then the parts can be computed simultaneously by different parts of the hardware.

The ultimate parallel computer is a *quantum computer*. Quantum computing pioneer David Deutsch controversially argues that "quantum computers share information with huge numbers of versions of themselves throughout the multiverse," and can get answers faster here in our Universe by in a sense getting help from these other versions.[4] We don't yet know whether a commercially competitive quantum computer can be built during the coming decades, because it depends both on whether quantum physics works as we think it does and on our ability to overcome daunting technical challenges, but companies and governments around the world are betting tens of millions of dollars annually on the possibility. Although quantum computers cannot speed up run-of-the-mill computations, clever algorithms have been developed that may dramatically speed up specific types of calculations, such as cracking cryptosystems and training neural networks. A quantum computer could also efficiently simulate the behavior of quantum-mechanical systems, including atoms, molecules and new materials, replacing measurements in chemistry labs in the same way that simulations on traditional computers have replaced measurements in wind tunnels.

What Is Learning?

Although a pocket calculator can crush me in an arithmetic contest, it will never improve its speed or accuracy, no matter how much it practices. It doesn't learn: for example, every time I press its square-root button, it computes exactly the same function in exactly the same way. Similarly, the first computer program that ever beat me at chess never learned from its mistakes, but merely implemented a function that its clever programmer had designed to compute a good next move. In contrast, when Magnus Carlsen lost his first game of chess at age five, he began a learning process that made him the World Chess Champion eighteen years later.

The ability to learn is arguably the most fascinating aspect of general intelligence. We've already seen how a seemingly dumb clump of matter can remember and compute, but how can it learn? We've seen that finding the answer to a difficult question corresponds to computing a function, and that appropriately arranged matter can calculate any computable function. When we humans first created pocket calculators and chess programs, *we* did the arranging. For matter to learn, it must instead rearrange *itself* to get better and better at computing the desired function—simply by obeying the laws of physics.

To demystify the learning process, let's first consider how a very simple physical system can learn the digits of π and other numbers. Above we saw how a surface with many valleys (see figure 2.3) can be used as a memory device: for example, if the bottom of one of the valleys is at position $x = \pi \approx 3.14159$ and there are no other valleys nearby, then you can put a ball at $x = 3$ and watch the system compute the missing decimals by letting the ball roll down to the bottom. Now, suppose that the surface is made of soft clay and starts out completely flat, as a blank slate. If some math enthusiasts repeatedly place the ball at the locations of each of their favorite numbers, then gravity will gradually create valleys at these locations, after which the clay surface can be used to recall these stored memories. In other words, the clay surface has *learned* to compute digits of numbers such as π.

Other physical systems, such as brains, can learn much more efficiently based on the same idea. John Hopfield showed that his above-mentioned network of interconnected neurons can learn in an analogous way: if you repeatedly put it into certain states, it will gradually learn these states and return to them from any nearby state. If you've seen each of your family members many times, then memories of what they look like can be triggered by anything related to them.

Neural networks have now transformed both biological and artificial intelligence, and have recently started dominating the AI subfield known as *machine learning* (the study of algorithms that improve through experience). Before delving deeper into how such networks can learn, let's first understand how they can compute. A neural network is simply a group of interconnected neurons that are able to influence each other's behavior. Your brain contains about as many neurons as there are stars in our Galaxy: in the ballpark of a hundred billion. On average, each of these neurons is connected to about a thousand others via junctions called *synapses*, and it's the strengths of these roughly hundred trillion synapse connections that encode most of the information in your brain.

We can schematically draw a neural network as a collection of dots representing neurons connected by lines representing synapses (see figure 2.9). Real-world neurons are very complicated electrochemical devices looking nothing like this schematic illustration: they involve different parts with names such as axons and dendrites, there are many different kinds of neurons that operate in a wide variety of ways, and the exact details of how and when electrical activity in one neuron affects other neurons is still the subject of active study. However, AI researchers have shown that neural networks can still attain human-level performance on many remarkably complex tasks even if one ignores all these complexities and replaces real biological neurons with extremely simple simulated ones that are all identical and obey very simple rules. The currently most popular model for such an *artificial neural network* represents the state of each neuron by a single number and the strength of each synapse by a single number. In this

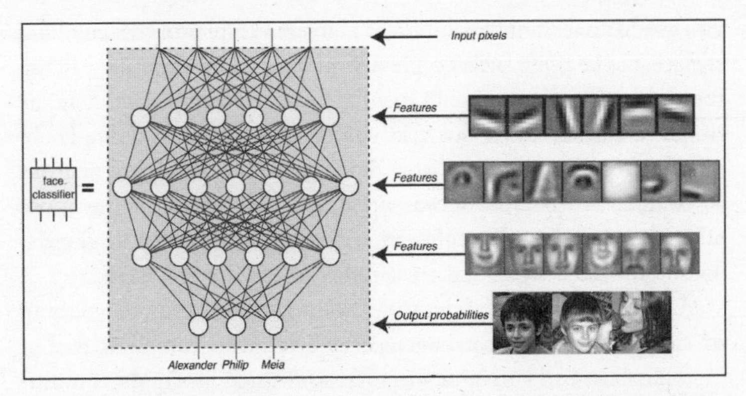

Figure 2.9: A network of neurons can compute functions just as a network of NAND gates can. For example, artificial neural networks have been trained to input numbers representing the brightness of different image pixels and output numbers representing the probability that the image depicts various people. Here each artificial neuron (circle) computes a weighted sum of the numbers sent to it via connections (lines) from above, applies a simple function and passes the result downward, each subsequent layer computing higher-level features. Typical face-recognition networks contain hundreds of thousands of neurons; the figure shows merely a handful for clarity.

model, each neuron updates its state at regular time steps by simply averaging together the inputs from all connected neurons, weighting them by the synaptic strengths, optionally adding a constant, and then applying what's called an *activation function* to the result to compute its next state.* The easiest way to use a neural network as a function is to make it *feedforward*, with information flowing only in one direction, as in figure 2.9, plugging the input to the function into a layer of neurons at the top and extracting the output from a layer of neurons at the bottom.

The success of these simple artificial neural networks is yet another example of substrate independence: neural networks have great computational power seemingly independent of the low-level nitty-gritty details of their construction. Indeed, George Cybenko, Kurt

* In case you like math, two popular choices of this activation function are the so-called sigmoid function $\sigma(x) \equiv 1/(1 + e^{-x})$ and the ramp function $\sigma(x) = \max\{0, x\}$, although it's been proven that almost any function will suffice as long as it's not linear (a straight line). Hopfield's famous model uses $\sigma(x) = -1$ if $x < 0$ and $\sigma(x) = 1$ if $x \geq 0$. If the neuron states are stored in a vector, then the network is updated by simply multiplying that vector by a matrix storing the synaptic couplings and then applying the function σ to all elements.

Hornik, Maxwell Stinchcombe and Halbert White proved something remarkable in 1989: such simple neural networks are *universal* in the sense that they can compute *any* function arbitrarily accurately, by simply adjusting those synapse strength numbers accordingly. In other words, evolution probably didn't make our biological neurons so complicated because it was necessary, but because it was more efficient—and because evolution, as opposed to human engineers, doesn't reward designs that are simple and easy to understand.

When I first learned about this, I was mystified by how something so simple could compute something arbitrarily complicated. For example, how can you compute even something as simple as multiplication, when all you're allowed to do is compute weighted sums and apply a single fixed function? In case you'd like a taste of how this works, figure 2.10 shows how a mere five neurons can multiply two arbitrary numbers together, and how a single neuron can multiply three bits together.

Although you can prove that you can compute anything in *theory* with an arbitrarily large neural network, the proof doesn't say anything about whether you can do so in *practice*, with a network of reasonable size. In fact, the more I thought about it, the more puzzled I became that neural networks worked so well.

For example, suppose that we wish to classify megapixel grayscale images into two categories, say cats or dogs. If each of the million pixels can take one of, say, 256 values, then there are $256^{1000000}$ possible images, and for each one, we wish to compute the probability that it depicts a cat. This means that an arbitrary function that inputs a picture and outputs a probability is defined by a list of $256^{1000000}$ probabilities, that is, way more numbers than there are atoms in our Universe (about 10^{78}). Yet neural networks with merely thousands or millions of parameters somehow manage to perform such classification tasks quite well. How can successful neural networks be "cheap," in the sense of requiring so few parameters? After all, you can prove that a neural network small enough to fit inside our Universe will epically fail to approximate almost all functions, succeeding merely on a ridiculously tiny fraction of all computational tasks that you might assign to it.

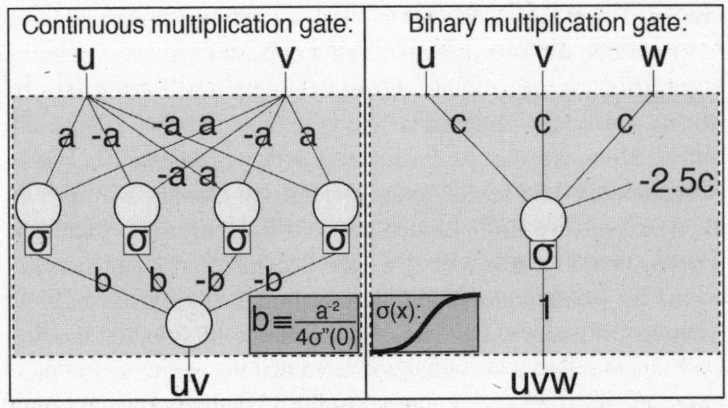

Figure 2.10: How matter can multiply, but using not NAND gates as in figure 2.7 but neurons. The key point doesn't require following the details, and is that not only can neurons (artificial or biological) do math, but multiplication requires many fewer neurons than NAND gates. *Optional details for hard-core math fans:* Circles perform summation, squares apply the function σ, and lines multiply by the constants labeling them. The inputs are real numbers (left) and bits (right). The multiplication becomes arbitrarily accurate as $a \to 0$ (left) and $c \to \infty$ (right). The left network works for any function $\sigma(x)$ that's curved at the origin (with second derivative $\sigma''(0) \neq 0$), which can be proven by Taylor expanding $\sigma(x)$. The right network requires that the function $\sigma(x)$ approaches 0 and 1 when x gets very small and very large, respectively, which is seen by noting that $uvw = 1$ only if $u + v + w = 3$. (These examples are from a paper I wrote with my students Henry Lin and David Rolnick, "Why Does Deep and Cheap Learning Work So Well?," which can be found at http://arxiv.org/abs/1608.08225.) By combining together lots of multiplications (as above) and additions, you can compute any polynomials, which are well known to be able to approximate any smooth function.

I've had lots of fun puzzling over this and related mysteries with my student Henry Lin. One of the things I feel most grateful for in life is the opportunity to collaborate with amazing students, and Henry is one of them. When he first walked into my office to ask whether I was interested in working with him, I thought to myself that it would be more appropriate for me to ask whether he was interested in working with me: this modest, friendly and bright-eyed kid from Shreveport, Louisiana, had already written eight scientific papers, won a Forbes 30-Under-30 award, and given a TED talk with over a million views—and he was only twenty! A year later, we wrote a paper together with a surprising conclusion: the question of why neural networks work so well can't be answered with mathematics alone, because part of the answer lies in physics. We found that the class of functions that the laws of physics throw at us and make us interested in computing is also a remarkably tiny class because, for reasons that we still don't fully understand, the laws of physics are remarkably simple. Moreover, the tiny fraction of functions that neural networks can compute is very similar to the tiny fraction that physics makes us interested in! We also extended previous work showing that deep-learning neural networks (they're called "deep" if they contain many layers) are much more efficient than shallow ones for many of these functions of interest. For example, together with another amazing MIT student, David Rolnick, we showed that the simple task of multiplying n numbers requires a whopping 2^n neurons for a network with only one layer, but takes only about $4n$ neurons in a deep network. This helps explain not only why neural networks are now all the rage among AI researchers, but also why we evolved neural networks in our brains: if we evolved brains to predict the future, then it makes sense that we'd evolve a computational architecture that's good at precisely those computational problems that matter in the physical world.

Now that we've explored how neural networks work and compute, let's return to the question of how they can learn. Specifically, how can a neural network get better at computing by updating its synapses?

In his seminal 1949 book, *The Organization of Behavior: A Neuropsychological Theory*, the Canadian psychologist Donald Hebb argued that if two nearby neurons were frequently active ("firing") at the same time, their synaptic coupling would strengthen so that they learned to help trigger each other—an idea captured by the popular slogan "Fire together, wire together." Although the details of how actual brains learn are still far from understood, and research has shown that the answers are in many cases much more complicated, it's also been shown that even this simple learning rule (known as Hebbian learning) allows neural networks to learn interesting things. John Hopfield showed that Hebbian learning allowed his oversimplified artificial neural network to store lots of complex memories by simply being exposed to them repeatedly. Such exposure to information to learn from is usually called "training" when referring to artificial neural networks (or to animals or people being taught skills), although "studying," "education" or "experience" might be just as apt. The artificial neural networks powering today's AI systems tend to replace Hebbian learning with more sophisticated learning rules with nerdy names such as "backpropagation" and "stochastic gradient descent," but the basic idea is the same: there's some simple deterministic rule, akin to a law of physics, by which the synapses get updated over time. As if by magic, this simple rule can make the neural network learn remarkably complex computations if training is performed with large amounts of data. We don't yet know precisely what learning rules our brains use, but whatever the answer may be, there's no indication that they violate the laws of physics.

Just as most digital computers gain efficiency by splitting their work into multiple steps and reusing computational modules many times, so do many artificial and biological neural networks. Brains have parts that are what computer scientists call *recurrent* rather than feedforward neural networks, where information can flow in multiple directions rather than just one way, so that the current output can become input to what happens next. The network of logic gates in the microprocessor of a laptop is also recurrent in this sense: it keeps reusing its past information, and lets new information input from a keyboard, trackpad, camera, etc., affect its ongoing computation, which in turn determines information output to, say, a screen, loudspeaker, printer or wireless network. Analogously, the network of neurons in your brain is recurrent, letting information input from your eyes, ears and other senses affect its ongoing computation, which in turn determines information output to your muscles.

The history of learning is at least as long as the history of life itself, since every self-reproducing organism performs interesting copying and processing of information—behavior that has somehow been learned. During the era of Life 1.0, however, organisms didn't learn during their lifetime: their rules for processing information and reacting were determined by their inherited DNA, so the only learning occurred slowly at the species level, through Darwinian evolution across generations.

About half a billion years ago, certain gene lines here on Earth discovered a way to make animals containing neural networks, able to learn behaviors from experiences during life. Life 2.0 had arrived, and because of its ability to learn dramatically faster and outsmart the competition, it spread like wildfire across the globe. As we explored in chapter 1, life has gotten progressively better at learning, and at an ever-increasing rate. A particular ape-like species grew a brain so adept at acquiring knowledge that it learned how to use tools, make fire, speak a language and create a complex global society. This society can itself be viewed as a system that remembers, computes and learns, all at an accelerating pace as one invention enables the next: writing, the printing press, modern science, computers, the internet and so on. What will future historians put next on that list of enabling inventions? My guess is artificial intelligence.

As we all know, the explosive improvements in computer memory and computational power (figure 2.4 and figure 2.8) have translated into spectacular progress in artificial intelligence—but it took a long time until machine *learning* came of age. When IBM's Deep Blue computer overpowered chess champion Garry Kasparov in 1997, its major advantages lay in memory and computation, not in learning. Its computational intelligence had been created by a team of humans, and the key reason that Deep Blue could outplay its creators was its ability to compute faster and thereby analyze more potential positions. When IBM's Watson computer dethroned the human world champion in the quiz show *Jeopardy!*, it too relied less on learning than on custom-programmed skills and superior memory and speed. The same can be said of most early breakthroughs in robotics, from legged locomotion to self-driving cars and self-landing rockets.

In contrast, the driving force behind many of the most recent AI breakthroughs has been machine *learning*. Consider figure 2.11, for example. It's easy for you to tell what it's a photo of, but to program a function that inputs nothing but the colors of all the pixels of an image and outputs an accurate caption such as "A group of young people playing a game of frisbee" had eluded all the world's AI researchers for decades. Yet a team at Google led by Ilya Sutskever did precisely that in 2014. Input a different set of pixel colors, and it replies "A herd of elephants walking across a dry grass field," again correctly. How did they do it? Deep Blue–style, by programming handcrafted algorithms for detecting frisbees, faces and the like? No, by creat-

Figure 2.11: "A group of young people playing a game of frisbee"—that caption was written by a computer with no understanding of people, games or frisbees.

ing a relatively simple neural network with no knowledge whatsoever about the physical world or its contents, and then letting it learn by exposing it to massive amounts of data. AI visionary Jeff Hawkins wrote in 2004 that "no computer can . . . see as well as a mouse," but those days are now long gone.

Just as we don't fully understand how our children learn, we still don't fully understand how such neural networks learn, and why they occasionally fail. But what's clear is that they're already highly useful and are triggering a surge of investments in deep learning. Deep learning has now transformed many aspects of computer vision, from handwriting transcription to real-time video analysis for self-driving cars. It has similarly revolutionized the ability of computers to transform spoken language into text and translate it into other languages, even in real time—which is why we can now talk to personal digital assistants such as Siri, Google Now and Cortana. Those annoying CAPTCHA puzzles, where we need to convince a website that we're human, are getting ever more difficult in order to keep ahead of what machine-learning technology can do. In 2015, Google DeepMind released an AI system using deep learning that was able to master dozens of computer games like a kid would—with no instructions whatsoever—except that it soon learned to play better than any human. In 2016, the same company built AlphaGo, a Go-playing computer system that used deep learning to evaluate the strength of different board positions and defeated the world's strongest Go champion. This progress is fueling a virtuous circle, bringing ever more funding and talent into AI research, which generates further progress.

We've spent this chapter exploring the nature of intelligence and its development up until now. How long will it take until machines can out-compete us at *all* cognitive tasks? We clearly don't know, and need to be open to the possibility that the answer may be "never." However, a basic message of this chapter is that we also need to consider the possibility that it *will* happen, perhaps even in our lifetime. After all, matter can be arranged so that when it obeys the laws of physics, it remembers, computes and learns—and the matter doesn't need to be biological. AI researchers have often been accused of over-promising and under-delivering, but in fairness, some of their critics don't have the best track record either. Some keep moving the goal-posts, effectively defining intelligence as that which computers still can't do, or as that which impresses us. Machines are now good or excellent at arithmetic, chess, mathematical theorem proving, stock picking, image captioning, driving, arcade game playing, Go, speech synthesis, speech transcription, translation and cancer diagnosis, but some critics will scornfully scoff "Sure—but that's not *real* intelligence!" They might go on to argue that real intelligence involves only the mountaintops in Moravec's landscape (figure 2.2) that haven't yet been submerged, just as some people in the past used to argue that image captioning and Go should count—while the water kept rising.

Assuming that the water will keep rising for at least a while longer, AI's impact on society will keep growing. Long before AI reaches human level across all tasks, it will give us fascinating opportunities and challenges involving issues such as bugs, laws, weapons and jobs. What are they and how can we best prepare for them? Let's explore this in the next chapter.

THE BOTTOM LINE:

- Intelligence, defined as ability to accomplish complex goals, can't be measured by a single IQ, only by an ability spectrum across all goals.
- Today's artificial intelligence tends to be *narrow*, with each system able to accomplish only very specific goals, while human intelligence is remarkably *broad*.
- Memory, computation, learning and intelligence have an abstract, intangible and ethereal feel to them because they're *substrate-independent:* able to take on a life of their own that doesn't depend on or reflect the details of their underlying material substrate.
- Any chunk of matter can be the substrate for *memory* as long as it has many different stable states.
- Any matter can be *computronium*, the substrate for *computation*, as long as it contains certain universal building blocks that can be combined to implement any function. NAND gates and neurons are two important examples of such universal "computational atoms."
- A neural network is a powerful substrate for *learning* because, simply by obeying the laws of physics, it can rearrange itself to get better and better at implementing desired computations.
- Because of the striking simplicity of the laws of physics, we humans only care about a tiny fraction of all imaginable computational problems, and neural networks tend to be remarkably good at solving precisely this tiny fraction.
- Once technology gets twice as powerful, it can often be used to design and build technology that's twice as powerful in turn, triggering repeated capability doubling in the spirit of Moore's law. The cost of information technology has now halved roughly every two years for about a century, enabling the information age.
- If AI progress continues, then long before AI reaches human level for all skills, it will give us fascinating opportunities and challenges involving issues such as bugs, laws, weapons and jobs—which we'll explore in the next chapter.

Chapter 3

The Near Future: Breakthroughs, Bugs, Laws, Weapons and Jobs

If we don't change direction soon, we'll end up where we're going.

Irwin Corey

What does it mean to be human in the present day and age? For example, what is it that we really value about ourselves, that makes us different from other life forms and machines? What do other people value about us that makes some of them willing to offer us jobs? Whatever our answers are to these questions at any one time, it's clear that the rise of technology must gradually change them.

Take me, for instance. As a scientist, I take pride in setting my own goals, in using creativity and intuition to tackle a broad range of unsolved problems, and in using language to share what I discover. Fortunately for me, society is willing to pay me to do this as a job. Centuries ago, I might instead, like many others, have built my identity around being a farmer or craftsman, but the growth of technology has since reduced such professions to a tiny fraction of the workforce. This means that it's no longer possible for everyone to build their identity around farming or crafts.

Personally, it doesn't bother me that today's machines outclass me at manual skills such as digging and knitting, since these are neither hobbies of mine nor my sources of income or self-worth. Indeed, any delusions I may have held about my abilities in that regard were crushed at age eight, when my school forced me to take a knitting class which I nearly flunked, and I completed my project only thanks to a compassionate helper from fifth grade taking pity on me.

But as technology keeps improving, will the rise of AI eventually eclipse also those abilities that provide my current sense of self-worth and value on the job market? Stuart Russell told me that he and many of his fellow AI researchers had recently experienced a "holy shit!" moment, when they witnessed AI doing something they weren't expecting to see for many years. In that spirit, please let me tell you about a few of my own HS moments, and how I see them as harbingers of human abilities soon to be overtaken.

Breakthroughs

Deep Reinforcement Learning Agents

I experienced one of my major jaw drops in 2014 while watching a video of a DeepMind AI system learning to play computer games. Specifically, the AI was playing Breakout (see figure 3.1), a classic Atari game I remember fondly from my teens. The goal is to maneuver a paddle so as to repeatedly bounce a ball off a brick wall; every time you hit a brick, it disappears and your score increases.

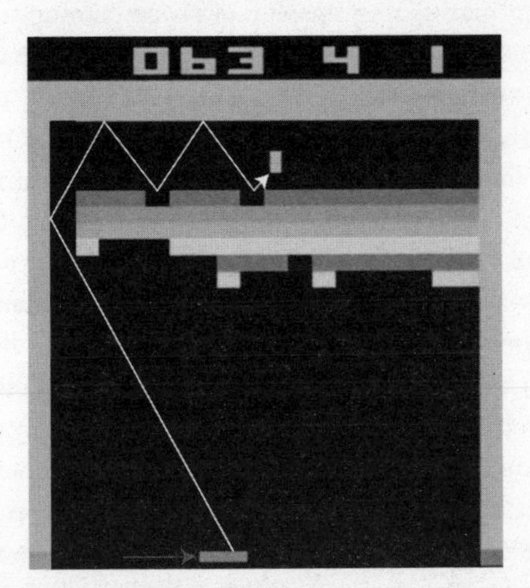

Figure 3.1: After learning to play the Atari game Breakout from scratch, using deep reinforcement learning to maximize the score, the DeepMind AI discovered the optimal strategy: drilling a hole through the leftmost part of the brick wall and letting the ball keep bouncing around behind it, amassing points very rapidly. I've drawn arrows showing the past trajectories of ball and paddle.

I'd written some computer games of my own back in the day, and was well aware that it wasn't hard to write a program that could play Breakout—but this was not what the DeepMind team had done. Instead, they'd created a blank-slate AI that knew nothing about this game—or about any other games, or even about *concepts* such as games, paddles, bricks or balls. All their AI knew was that a long list of numbers got fed into it at regular intervals: the current score and a long list of numbers which we (but not the AI) would recognize as specifications of how different parts of the screen were colored. The AI was simply told to maximize the score by outputting, at regular intervals, numbers which we (but not the AI) would recognize as codes for which keys to press.

Initially, the AI played terribly: it cluelessly jiggled the paddle back and forth seemingly at random and missed the ball almost every time. After a while, it seemed to be getting the idea that moving the paddle toward the ball was a good idea, even though it still missed most of the time. But it kept improving with practice, and soon got better at the game than I'd ever been, infallibly returning the ball no matter how fast it approached. And then my jaw dropped: it figured out this amazing score-maximizing strategy of always aiming for the upper-left corner to drill a hole through the wall and let the ball get stuck bouncing between the back of the wall and the barrier behind it. This felt like a really intelligent thing to do. Indeed, Demis Hassabis later told me that the programmers on that DeepMind team didn't know this trick until they learned it from the AI they'd built. I recommend watching a video of this for yourself at the link I've provided.[1]

There was a human-like feature to this that I found somewhat unsettling: I was watching an AI that had a goal and learned to get ever better at achieving it, eventually outperforming its creators. In the previous chapter, we defined intelligence as simply the ability to accomplish complex goals, so in this sense, DeepMind's AI was growing more intelligent in front of my eyes (albeit merely in the very narrow sense of playing this particular game). In the first chapter, we encountered what computer scientists call *intelligent agents:* entities that collect information about their environment from sensors and then process this information to decide how to act back on their environment. Although DeepMind's game-playing AI lived in an extremely simple virtual world composed of bricks, paddles and balls, I couldn't deny that it was an intelligent agent.

DeepMind soon published their method and shared their code, explaining that it used a very simple yet powerful idea called *deep reinforcement learning*.[2] Basic reinforcement learning is a classic machine learning technique inspired by behaviorist psychology, where getting a positive reward increases your tendency to do something again and vice versa. Just like a dog learns to do tricks when this increases the likelihood of its getting encouragement or a snack from its owner soon, DeepMind's AI learned to move the paddle to catch the ball because this increased the likelihood of its getting more points soon. DeepMind combined this idea with deep learning: they trained a deep neural net, as in the previous chapter, to predict how many points would on average be gained by pressing each of the allowed keys on the keyboard, and then the AI selected whatever key the neural net rated as most promising given the current state of the game.

When I listed traits contributing to my own personal feeling of self-worth as a human, I included the ability to tackle a broad range of unsolved problems. In contrast, being able to play Breakout and do nothing else constitutes extremely narrow intelligence. To me, the true importance of DeepMind's breakthrough is that deep reinforcement learning is a completely general technique. Sure enough, they let the exact same AI practice playing forty-nine different Atari games, and it learned to outplay their human testers on twenty-nine of them, from Pong to Boxing, Video Pinball and Space Invaders.

It didn't take long until the same AI idea had started proving itself on more modern games whose worlds were three-dimensional rather than two-dimensional. Soon DeepMind's San Francisco–based competitors at OpenAI released a platform called Universe, where DeepMind's AI and other intelligent agents can practice interacting with an entire computer as if it were a game: clicking on anything, typing anything, and opening and running whatever software they're able to navigate—firing up a web browser and messing around online, for example.

Looking to the future of deep reinforcement learning and improvements thereupon, there's no obvious end in sight. The potential isn't limited to virtual game worlds, since if you're a robot, life itself can be viewed as a game. Stuart Russell told me that his first major HS moment was watching the robot Big Dog run up a snow-covered forest slope, elegantly solving the legged locomotion problem that he himself had struggled to solve for many years.[3] Yet when that milestone was reached in 2008, it involved huge amounts of work by clever programmers. After DeepMind's breakthrough, there's no reason why a robot can't ultimately use some variant of deep reinforcement learning to teach itself to walk without help from human programmers: all that's needed is a system that gives it points whenever it makes progress. Robots in the real world similarly have the potential to learn to swim, fly, play ping-pong, fight and perform a nearly endless list of other motor tasks without help from human programmers. To speed things up and reduce the risk of getting stuck or damaging themselves during the learning process, they would probably do the first stages of their learning in virtual reality.

Intuition, Creativity and Strategy

Another defining moment for me was when the DeepMind AI system AlphaGo won a five-game Go match against Lee Sedol, generally considered the top player in the world in the early twenty-first century.

It was widely expected that human Go players would be dethroned by machines at some point, since it had happened to their chess-playing colleagues two decades earlier. However, most Go pundits predicted that it would take another decade, so AlphaGo's triumph was a pivotal moment for them as well as for me. Nick Bostrom and Ray Kurzweil have both emphasized how hard it can be to see AI breakthroughs coming, which is evident from interviews with Lee Sedol himself before and after losing the first three games:

- October 2015: "Based on its level seen . . . I think I will win the game by a near landslide."
- February 2016: "I have heard that Google DeepMind's AI is surprisingly strong and getting stronger, but I am confident that I can win at least this time."

- March 9, 2016: "I was very surprised because I didn't think I would lose."
- March 10, 2016: "I'm quite speechless . . . I am in shock. I can admit that . . . the third game is not going to be easy for me."
- March 12, 2016: "I kind of felt powerless."

Within a year after playing Lee Sedol, a further improved AlphaGo had played all twenty top players in the world without losing a single match.

Why was this such a big deal for me personally? Well, I confessed above that I view intuition and creativity as two of my core human traits, and as I'll now explain, I feel that AlphaGo displayed both.

Go players take turns placing black and white stones on a 19-by-19 board (see figure 3.2). There are vastly more possible Go positions than there are atoms in our Universe, which means that trying to analyze all interesting sequences of future moves rapidly gets hopeless. Players therefore rely heavily on subconscious intuition to complement their conscious reasoning, with experts developing an almost uncanny feel for which positions are strong and which are weak. As we saw in the last chapter, the results of deep learning are sometimes

Figure 3.2: DeepMind's AlphaGo AI made a highly creative move on line 5, in defiance of millennia of human wisdom, which about fifty moves later proved crucial to its defeat of Go legend Lee Sedol.

reminiscent of intuition: a deep neural network might determine that an image portrays a cat without being able to explain why. The DeepMind team therefore gambled on the idea that deep learning might be able to recognize not merely cats, but also strong Go positions. The core idea that they built into AlphaGo was to marry the intuitive power of deep learning with the logical power of GOFAI—which stands for what's humorously known as "Good Old-Fashioned AI" from before the deep-learning revolution. They used a massive database of Go positions from both human play and games where AlphaGo had played a clone of itself, and trained a deep neural network to predict from each position the probability that white would ultimately win. They also trained a separate network to predict likely next moves. They then combined these networks with a GOFAI method that cleverly searched through a pruned list of likely future-move sequences to identify the next move that would lead to the strongest position down the road.

This marriage of intuition and logic gave birth to moves that were not merely powerful, but in some cases also highly creative. For example, millennia of Go wisdom dictate that early in the game, it's best to play on the third or fourth line from an edge. There's a trade-off between the two: playing on the third line helps with short-term territory gain toward the side of the board, while playing on the fourth helps with long-term strategic influence toward the center.

In the thirty-seventh move of the second game, AlphaGo shocked the Go world by defying that ancient wisdom and playing on the fifth line (figure 3.2), as if it were even more confident than a human in its long-term planning abilities and therefore favored strategic advantage over short-term gain. Commentators were stunned, and Lee Sedol even got up and temporarily left the room.[4] Sure enough, about fifty moves later, fighting from the lower left-hand corner of the board ended up spilling over and connecting with that black stone from move thirty-seven! And that motif is what ultimately won the game, cementing the legacy of AlphaGo's fifth-row move as one of the most creative in Go history.

Because of its intuitive and creative aspects, Go is viewed more as an art form than just another game. It was considered one of the four "essential arts" in ancient China, together with painting, calligraphy and *qin* music, and it remains hugely popular in Asia, with almost 300 million people watching the first game between AlphaGo and Lee Sedol. As a result, the Go world was quite shaken by the outcome, and viewed AlphaGo's victory as a profound milestone for humanity. Ke Jie, the world's top-ranked Go player at the time, had this to say:[5] "Humanity has played Go for thousands of years, and yet, as AI has shown us, we have not yet even scratched the surface . . . The union of human and computer players will usher in a new era . . . Together, man and AI can find the truth of Go." Such fruitful human-machine collaboration indeed appears promising in many areas, including science, where AI can hopefully help us humans deepen our understanding and realize our ultimate potential.

To me, AlphaGo also teaches us another important lesson for the near future: combining the intuition of deep learning with the logic of GOFAI can produce second-to-none *strategy*. Because Go is one of the ultimate strategy games, AI is now poised to graduate and challenge (or help) the best human strategists even beyond game boards—for example with investment strategy, political strategy and military strategy. Such real-world strategy problems are typically complicated by human psychology, missing information and factors that need to be modeled as random, but poker-playing AI systems have already demonstrated that none of these challenges are insurmountable.

Natural Language

Yet another area where AI progress has recently stunned me is language. I fell in love with travel early in life, and curiosity about other cultures and languages formed an important part of my identity. I was raised speaking Swedish and English, was taught German and Spanish in school, learned Portuguese and Romanian through two marriages and taught myself some Russian, French and Mandarin for fun.

But the AI has been reaching, and after an important discovery in 2016, there are almost no lazy languages that I can translate between better than the system of the AI developed by the equipment of the brain of Google.

Did I make myself crystal clear? I was actually trying to say this:

But AI has been catching up with me, and after a major breakthrough in 2016, there are almost no languages left that I can translate between better than the AI system developed by the Google Brain team.

However, I first translated it to Spanish and back using an app that I installed on my laptop a few years ago. In 2016, the Google Brain team upgraded their free Google Translate service to use deep recurrent neural networks, and the improvement over older GOFAI systems was dramatic:[6]

But AI has been catching up on me, and after a breakthrough in 2016, there are almost no languages left that can translate between better than the AI system developed by the Google Brain team.

As you can see, the pronoun "I" got lost during the Spanish detour, which unfortunately changed the meaning. Close, but no cigar! However, in defense of Google's AI, I'm often criticized for writing unnecessarily long sentences that are hard to parse, and I picked one of my most confusingly convoluted ones for this example. For more typical sentences, their AI often translates impeccably. As a result, it created quite a stir when it came out, and it's helpful enough to be used by hundreds of millions of people daily. Moreover, courtesy of recent progress in deep learning for speech-to-text and text-to-speech conversion, these users can now speak to their smartphones in one language and listen to the translated result.

Natural language processing is now one of the most rapidly advancing fields of AI, and I think that further success will have a large impact because language is so central to being human. The better an AI gets at linguistic prediction, the better it can compose reasonable email responses or continue a spoken conversation. This might, at least to an outsider, give the appearance of human thought taking place. Deep-learning systems are thus taking baby steps toward passing the famous Turing test, where a machine has to converse well enough in writing to trick a person into thinking that it too is human.

Language-processing AI still has a long way to go, though. Although I must confess that I feel a bit deflated when I'm out-translated by an AI, I feel better once I remind myself that, so far, it doesn't *understand* what it's saying in any meaningful sense. From being trained on massive data sets, it discovers patterns and relations involving words without ever relating these words to anything in the real world. For example, it might represent each word by a list of a thousand numbers that specify how similar it is to certain other words. It may then conclude from this that the difference between "king" and "queen" is similar to that between "husband" and "wife"—but it still has no clue what it means to be male or female, or even that there is such a thing as a physical reality out there with space, time and matter.

Since the Turing test is fundamentally about deception, it has been criticized for testing human gullibility more than true artificial intelligence. In contrast, a rival test called the *Winograd Schema Challenge* goes straight for the jugular, homing in on that commonsense understanding that current deep-learning systems tend to lack. We humans routinely use real-world knowledge when parsing a sentence, to figure out what a pronoun refers to. For example, a typical Winograd challenge asks what "they" refers to here:

1. "The city councilmen refused the demonstrators a permit because they feared violence."
2. "The city councilmen refused the demonstrators a permit because they advocated violence."

There's an annual AI competition to answer such questions, and AIs still perform miserably.[7] This precise challenge, understanding what refers to what, torpedoed even GoogleTranslate when I replaced Spanish with Chinese in my example above:

But the AI has caught up with me, after a major break in 2016, with almost no language, I could translate the AI system than developed by the Google Brain team.

Please try it yourself at https://translate.google.com now that you're reading the book and see if Google's AI has improved! There's a good chance that it has, since there are promising approaches out there for marrying deep recurrent neural nets with GOFAI to build a language-processing AI that includes a world model.

Opportunities and Challenges

These three examples were obviously just a sampler, since AI is progressing rapidly across many important fronts. Moreover, although I've mentioned only two companies in these examples, competing research groups at universities and other companies often weren't far behind. A loud sucking noise can be heard in computer science departments around the world as Apple, Baidu, DeepMind, Facebook, Google, Microsoft and others use lucrative offers to vacuum off students, postdocs and faculty.

It's important not to be misled by the examples I've given into viewing the history of AI as periods of stagnation punctuated by the occasional breakthrough. From my vantage point, I've instead been seeing fairly steady progress for a long time—which the media report as a breakthrough whenever it crosses the threshold of enabling a new imagination-grabbing application or useful product. I therefore consider it likely that brisk AI progress will continue for many years. Moreover, as we saw in the last chapter, there's no fundamental reason why this progress can't continue until AI matches human abilities on most tasks.

Which raises the question: How will this impact us? How will near-term AI progress change what it means to be human? We've seen that it's getting progressively harder to argue that AI completely lacks goals, breadth, intuition, creativity or language—traits that many feel are central to being human. This means that even in the near term, long before any AGI can match us at all tasks, AI might have a dramatic impact on how we view ourselves, on what we can do when complemented by AI and on what we can earn money doing when competing against AI. Will this impact be for the better or for the worse? What near-term opportunities and challenges will this present?

Everything we love about civilization is the product of human intelligence, so if we can amplify it with artificial intelligence, we obviously have the potential to make life even better. Even modest progress in AI might translate into major improvements in science and technology and corresponding reductions of accidents, disease, injustice, war, drudgery and poverty. But in order to reap these benefits of AI without creating new problems, we need to answer many important questions. For example:

1. How can we make future AI systems more robust than today's, so that they do what we want without crashing, malfunctioning or getting hacked?

2. How can we update our legal systems to be more fair and efficient and to keep pace with the rapidly changing digital landscape?
3. How can we make weapons smarter and less prone to killing innocent civilians without triggering an out-of-control arms race in lethal autonomous weapons?
4. How can we grow our prosperity through automation without leaving people lacking income or purpose?

Let's devote the rest of this chapter to exploring each of these questions in turn. These four near-term questions are aimed mainly at computer scientists, legal scholars, military strategists and economists, respectively. However, to help get the answers we need by the time we need them, everybody needs to join this conversation, because as we'll see, the challenges transcend all traditional boundaries—both between specialties and between nations.

Bugs vs. Robust AI

Information technology has already had great positive impact on virtually every sector of our human enterprise, from science to finance, manufacturing, transportation, healthcare, energy and communication, and this impact pales in comparison to the progress that AI has the potential to bring. But the more we come to rely on technology, the more important it becomes that it's robust and trustworthy, doing what we want it to do.

Throughout human history, we've relied on the same tried-and-true approach to keeping our technology beneficial: learning from mistakes. We invented fire, repeatedly messed up, and then invented the fire extinguisher, fire exit, fire alarm and fire department. We invented the automobile, repeatedly crashed, and then invented seat belts, air bags and self-driving cars. Up until now, our technologies have typically caused sufficiently few and limited accidents for their harm to be outweighed by their benefits. As we inexorably develop ever more powerful technology, however, we'll inevitably reach a point where even a single accident could be devastating enough to outweigh all benefits. Some argue that accidental global nuclear war would constitute such an example. Others argue that a bioengineered pandemic could qualify, and in the next chapter, we'll explore the controversy around whether future AI could cause human extinction. But we need not consider such extreme examples to reach a crucial conclusion: as technology grows more powerful, we should rely less on the trial-and-error approach to safety engineering. In other words, *we should become more proactive than reactive*, investing in safety research aimed at preventing accidents from happening even once. This is why society invests more in nuclear-reactor safety than mousetrap safety.

This is also the reason why, as we saw in chapter 1, there was strong community interest in AI-safety research at the Puerto Rico conference. Computers and AI systems have always crashed, but this time is different: AI is gradually entering the real world, and it's not merely a nuisance if it crashes the power grid, the stock market or a nuclear weapons system. In the rest of this section, I want to introduce you to the four main areas of technical AI-safety research that are dominating the current AI-safety discussion and that are being pursued around the world: *verification*, *validation*, *security* and *control*.* To prevent things from getting too nerdy and dry, let's do this by exploring past successes and failures of information technology in different areas, as well as valuable lessons we can learn from them and research challenges that they pose.

Although most of these stories are old, involving low-tech computer systems that almost nobody would refer to as AI and that caused few, if any, casualties, we'll see that they nonetheless teach us valuable lessons for designing safe and powerful future AI systems whose failures could be truly catastrophic.

AI for Space Exploration

Let's start with something close to my heart: space exploration. Computer technology has enabled us to fly people to the Moon and to

* If you want a more detailed map of the AI-safety research landscape, there's an interactive one here, developed in a community effort spearheaded by FLI's Richard Mallah: https://futureoflife.org/landscape.

send unmanned spacecraft to explore all the planets of our Solar System, even landing on Saturn's moon Titan and on a comet. As we'll explore in chapter 6, future AI may help us explore other solar systems and galaxies—if it's bug-free. On June 4, 1996, scientists hoping to research Earth's magnetosphere cheered jubilantly as an Ariane 5 rocket from the European Space Agency roared into the sky with the scientific instruments they'd built. Thirty-seven seconds later, their smiles vanished as the rocket exploded in a fireworks display costing hundreds of millions of dollars.[8] The cause was found to be buggy software manipulating a number that was too large to fit into the 16 bits allocated for it.[9] Two years later, NASA's Mars Climate Orbiter accidentally entered the Red Planet's atmosphere and disintegrated because two different parts of the software used different units for force, causing a 445% error in the rocket-engine thrust control.[10] This was NASA's second super-expensive bug: their Mariner 1 mission to Venus exploded after launch from Cape Canaveral on July 22, 1962, after the flight-control software was foiled by an incorrect punctuation mark.[11] As if to show that not only westerners had mastered the art of launching bugs into space, the Soviet Phobos 1 mission failed on September 2, 1988. This was the heaviest interplanetary spacecraft ever launched, with the spectacular goal of deploying a lander on Mars' moon Phobos—all thwarted when a missing hyphen caused the "end-of-mission" command to be sent to the spacecraft while it was en route to Mars, shutting down all of its systems.[12]

What we learn from these examples is the importance of what computer scientists call *verification:* ensuring that software fully satisfies all the expected requirements. The more lives and resources are at stake, the higher confidence we want that the software will work as intended. Fortunately, AI can help automate and improve the verification process. For example, a complete, general-purpose operating-system kernel called *seL4* has recently been mathematically checked against a formal specification to give a strong guarantee against crashes and unsafe operations: although it doesn't yet come with the bells and whistles of Microsoft Windows and Mac OS, you can rest assured that it won't give you what's affectionately known as "the blue screen of death" or "the spinning wheel of doom." The U.S.

Defense Advanced Research Projects Agency (DARPA) has sponsored the development of a set of open-source high-assurance tools called HACMS (high-assurance cyber military systems) that are provably safe. An important challenge is to make such tools sufficiently powerful and easy to use that they'll get widely deployed. Another challenge is that the very task of verification will itself get more difficult as software moves into robots and new environments, and as traditional preprogrammed software gets replaced by AI systems that keep learning, thereby changing their behavior, as in chapter 2.

AI for Finance

Finance is another area that's been transformed by information technology, allowing resources to be efficiently reallocated across the globe at the speed of light and enabling affordable financing for everything from mortgages to startup companies. Progress in AI is likely to offer great future profit opportunities from financial trading: most stock market buy/sell decisions are now made automatically by computers, and my graduating MIT students routinely get tempted by astronomical starting salaries to improve algorithmic trading.

Verification is important for financial software as well, which the American firm Knight Capital learned the hard way on August 1, 2012, by losing $440 million in forty-five minutes after deploying unverified trading software.[13] The trillion-dollar "Flash Crash" of May 6, 2010, was noteworthy for a different reason. Although it caused massive disruptions for about half an hour before markets stabilized, with shares of some prominent companies such Procter & Gamble swinging in price between a penny and $100,000,[14] the problem wasn't caused by bugs or computer malfunctions that verification could have avoided. Instead, it was caused by expectations being violated: automatic trading programs from many companies found themselves operating in an unexpected situation where their assumptions weren't valid—for example, the assumption that if a stock exchange computer reported that a stock had a price of one cent, then that stock really was worth one cent.

The flash crash illustrates the importance of what computer scientists call *validation:* whereas verification asks "Did I build the system right?," validation asks "Did I build the right system?"* For example, does the system rely on assumptions that might not always be valid? If so, how can it be improved to better handle uncertainty?

AI for Manufacturing

Needless to say, AI holds great potential for improving manufacturing, by controlling robots that enhance both efficiency and precision. Ever-improving 3-D printers can now make prototypes of anything from office buildings to micromechanical devices smaller than a salt grain.[15] While huge industrial robots build cars and airplanes, affordable computer-controlled mills, lathes, cutters and the like are powering not merely factories, but also the grassroots "maker movement," where local enthusiasts materialize their ideas at over a thousand community-run "fab labs" around the world.[16] But the more robots we have around us, the more important it becomes that we verify and validate their software. The first person known to have been killed by a robot was Robert Williams, a worker at a Ford plant in Flat Rock, Michigan. In 1979, a robot that was supposed to retrieve parts from a storage area malfunctioned, and he climbed into the area to get the parts himself. The robot silently began operating and smashed his head, continuing for thirty minutes until his co-workers discovered what had happened.[17] The next robot victim was Kenji Urada, a maintenance engineer at a Kawasaki plant in Akashi, Japan. While working on a broken robot in 1981, he accidentally hit its on switch and was crushed to death by the robot's hydraulic arm.[18] In 2015, a twenty-two-year-old contractor at one of Volkswagen's production plants in Baunatal, Germany, was working on setting up a robot to grab auto parts and manipulate them. Something went wrong, causing the robot to grab him and crush him to death against a metal plate.[19]

Although these accidents are tragic, it's important to note that they make up a minuscule fraction of all industrial accidents. Moreover, industrial accidents have *decreased* rather than increased as technology

* More precisely, verification asks if a system meets its specifications, whereas validation asks if the correct specifications were chosen.

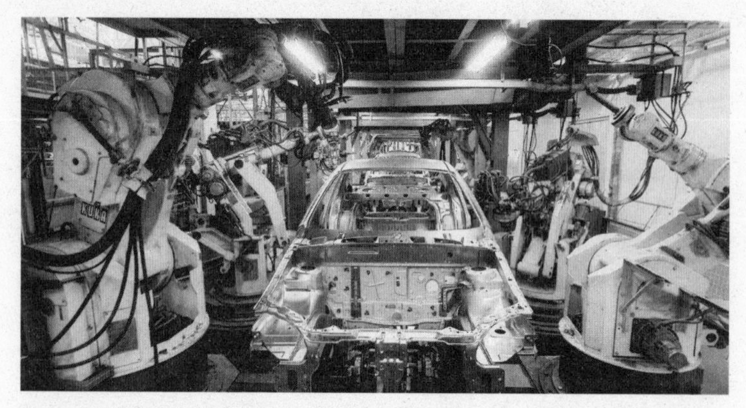

Figure 3.3: Whereas traditional industrial robots are expensive and hard to program, there's a trend toward cheaper AI-powered ones that can learn what to do from workers with no programming experience.

has improved, dropping from about 14,000 deaths in 1970 to 4,821 in 2014 in the United States.[20] The three above-mentioned accidents show that adding intelligence to otherwise dumb machines should be able to further improve industrial safety, by having robots learn to be more careful around people. All three accidents could have been avoided with better validation: the robots caused harm not because of bugs or malice, but because they made invalid assumptions—that the person wasn't present or that the person was an auto part.

AI for Transportation

Although AI can save many lives in manufacturing, it can potentially save even more in transportation. Car accidents alone took over 1.2 million lives in 2015, and aircraft, train and boat accidents together killed thousands more. In the United States, with its high safety standards, motor vehicle accidents killed about 35,000 people last year—seven times more than all industrial accidents combined.[21] When we had a panel discussion about this in Austin, Texas, at the 2016 annual meeting of the Association for the Advancement of Artificial Intelligence, the Israeli computer scientist Moshe Vardi got quite emotional about it and argued that not only *could* AI reduce road fatalities, but it *must:* "It's a moral imperative!" he exclaimed. Because almost all car crashes are caused by human error, it's widely believed that AI-powered self-driving cars can eliminate at least 90% of road deaths, and this optimism is fueling great progress toward actually getting self-driving cars out on the roads. Elon Musk envisions that future self-driving cars will not only be safer, but will also earn money for their owners while they're not needed, by competing with Uber and Lyft.

So far, self-driving cars do indeed have a better safety record than human drivers, and the accidents that have occurred underscore the importance and difficulty of validation. The first fender bender caused by a Google self-driving car took place on February 14, 2016, because it made an incorrect assumption about a bus: that its driver would yield when the car pulled out in front of it. The first lethal crash caused by a self-driving Tesla, which rammed into the trailer of a truck crossing the highway on May 7, 2016, was caused by two bad assumptions:[22] that the bright white side of the trailer was merely part of the bright sky, and that the driver (who was allegedly watching a Harry Potter movie) was paying attention and would intervene if something went wrong.*

But sometimes good verification and validation aren't enough to avoid accidents, because we also need good *control:* ability for a human operator to monitor the system and change its behavior if necessary. For such *human-in-the-loop* systems to work well, it's crucial that the human-machine communication be effective. In this spirit, a red light on your dashboard will conveniently alert you if you accidentally leave the trunk of your car open. In contrast, when the British car ferry *Herald of Free Enterprise* left the harbor of Zeebrugge on March 6, 1987, with her bow doors open, there was no warning light or other visible warning for the captain, and the ferry capsized soon after leaving the harbor, killing 193 people.[23]

Another tragic control failure that might have been avoided by better machine-human communication occurred during the night of June 1, 2009, when Air France Flight 447 crashed into the Atlan-

* Even including this crash in the statistics, Tesla's Autopilot was found to reduce crashes by 40% when turned on: http://tinyurl.com/teslasafety.

tic Ocean, killing all 228 on board. According to the official accident report, "the crew never understood that they were stalling and consequently never applied a recovery manoeuvre"—which would have involved pushing down the nose of the aircraft—until it was too late. Flight safety experts speculated that the crash might have been avoided had there been an "angle-of-attack" indicator in the cockpit, showing the pilots that the nose was pointed too far upward.[24]

When Air Inter Flight 148 crashed into the Vosges Mountains near Strasbourg in France on January 20, 1992, killing 87 people, the cause wasn't lack of machine-human communication, but a confusing user interface. The pilots entered "33" on a keypad because they wanted to descend at an angle of 3.3 degrees, but the autopilot interpreted this as 3,300 feet per minute because it was in a different mode—and the display screen was too small to show the mode and allow the pilots to realize their mistake.

AI for Energy

Information technology has done wonders for power generation and distribution, with sophisticated algorithms balancing production and consumption across the world's electrical grids, and sophisticated control systems keeping power plants operating safely and efficiently. Future AI progress is likely to make the "smart grid" even smarter, to optimally adapt to changing supply and demand even down to the level of individual rooftop solar panels and home-battery systems. But on Thursday, August 14, 2003, it was lights-out for about 55 million people in the United States and Canada, many of whom remained powerless for days. Here, too, the primary cause was determined to be failed machine-human communications: a software bug prevented the alarm system in an Ohio control room from alerting operators to the need to redistribute power before a minor problem (overloaded transmission lines hitting unpruned foliage) cascaded out of control.[25]

The partial nuclear meltdown in a reactor on Three Mile Island in Pennsylvania on March 28, 1979, led to about a billion dollars in cleanup cost and a major backlash against nuclear power. The final accident report identified multiple contributing factors, including confusion caused by a poor user interface.[26] In particular, the warning light that the operators thought indicated whether a safety-critical valve was open or closed merely indicated whether a signal had been sent to close the valve—so the operators didn't realize that the valve had gotten stuck open.

These energy and transportation accidents teach us that as we put AI in charge of ever more physical systems, we need to put serious research efforts into not only making the machines work well on their own, but also into making machines collaborate effectively with their human controllers. As AI gets smarter, this will involve not merely building good user interfaces for information sharing, but also figuring out how to optimally allocate tasks within human-computer teams—for example, identifying situations where control should be transferred, and for applying human judgment efficiently to the highest-value decisions rather than distracting human controllers with a flood of unimportant information.

AI for Healthcare

AI has huge potential for improving healthcare. Digitization of medical records has already enabled doctors and patients to make faster and better decisions, and to get instant help from experts around the world in diagnosing digital images. Indeed, the best experts for performing such diagnosis may soon be AI systems, given the rapid progress in computer vision and deep learning. For example, a 2015 Dutch study showed that computer diagnosis of prostate cancer using magnetic resonance imaging (MRI) was as good as that of human radiologists,[27] and a 2016 Stanford study showed that AI could diagnose lung cancer using microscope images even better than human pathologists.[28] If machine learning can help reveal relationships between genes, diseases and treatment responses, it could revolutionize personalized medicine, make farm animals healthier and enable more resilient crops. Moreover, robots have the potential to become more accurate and reliable surgeons than humans, even without using advanced AI. A wide variety of robotic surgeries have been successfully performed in recent years, often allowing precision, miniaturization and smaller incisions that lead to decreased blood loss, less pain and shorter healing time.

Alas, there have been painful lessons about the importance of robust software also in the healthcare industry. For example, the Canadian-built Therac-25 radiation therapy machine was designed to treat cancer patients in two different modes: either with a low-power beam of electrons or with a high-power beam of megavolt X-rays that was kept on target by a special shield. Unfortunately, unverified buggy software occasionally caused technicians to deliver the megavolt beam when they thought they were administering the low-power beam, and without the shield, which ended up claiming the lives of several patients.[29] Many more patients died from radiation overdoses at the National Oncologic Institute in Panama, where radiotherapy equipment using radioactive cobalt-60 was programmed to excessive exposure times in 2000 and 2001 because of a confusing user interface that hadn't been properly validated.[30] According to a recent report,[31] robotic surgery accidents were linked to 144 deaths and 1,391 injuries in the United States between 2000 and 2013, with common problems including not only hardware issues such as electrical arcing and burnt or broken pieces of instruments falling into the patient, but also software problems such as uncontrolled movements and spontaneous powering-off.

The good news is that the rest of almost two million robotic surgeries covered by the report went smoothly, and robots appear to be making surgery more rather than less safe. According to a U.S. government study, bad hospital care contributes to over 100,000 deaths per year in the United States alone,[32] so the moral imperative for developing better AI for medicine is arguably even stronger than that for self-driving cars.

AI for Communication

The communication industry is arguably the one where computers have had the greatest impact of all so far. After the introduction of computerized telephone switchboards in the fifties, the internet in the sixties, and the World Wide Web in 1989, billions of people now go online to communicate, shop, read news, watch movies or play games, accustomed to having the world's information just a click away—and often for free. The emerging *internet of things* promises improved efficiency, accuracy, convenience and economic benefit from bringing online everything from lamps, thermostats and freezers to biochip transponders on farm animals.

These spectacular successes in connecting the world have brought computer scientists a fourth challenge: they need to improve not only verification, validation and control, but also *security* against malicious software ("malware") and hacks. Whereas the aforementioned problems all resulted from unintentional mistakes, security is directed at *deliberate malfeasance*. The first malware to draw significant media attention was the so-called Morris worm, unleashed on November 2, 1988, which exploited bugs in the UNIX operating system. It was allegedly a misguided attempt to count how many computers were online, and although it infected and crashed about 10% of the 60,000 computers that made up the internet back then, this didn't stop its creator, Robert Morris, from eventually getting a tenured professorship in computer science at MIT.

Other malware exploits vulnerabilities not in software but in people. On May 5, 2000, as if to celebrate my birthday, people got emails with the subject line "ILOVEYOU" from acquaintances and colleagues, and those Microsoft Windows users who clicked on the attachment "LOVE-LETTER-FOR-YOU.txt.vbs" unwittingly launched a script that damaged their computer and re-sent the email to everyone in their address book. Created by two young programmers in the Philippines, this worm infected about 10% of the internet, just as the Morris worm had done, but because the internet was a lot bigger by then, it became one of the greatest infections of all time, afflicting over 50 million computers and causing over $5 billion in damages. As you're probably painfully aware, the internet remains infested with countless kinds of infectious malware, which security experts classify into worms, Trojans, viruses and other intimidating-sounding categories, and the damage they cause ranges from displaying harmless prank messages to deleting your files, stealing your personal information, spying on you and hijacking your computer to send out spam.

Whereas malware targets whatever computer it can, *hackers* attack specific targets of interest—recent high-profile examples including

Target, TJ Maxx, Sony Pictures, Ashley Madison, the Saudi oil company Aramco and the U.S. Democratic National Committee. Moreover, the loots appear to be getting ever more spectacular. Hackers stole 130 million credit card numbers and other account information from Heartland Payment Systems in 2008, and breached over a billion(!) Yahoo! email accounts in 2013.[33] A 2014 hack of the U.S. Government's Office of Personnel Management breached personnel records and job application information for over 21 million people, allegedly including employees with top security clearances and the fingerprints of undercover agents.

As a result, I roll my eyes whenever I read about some new system being allegedly 100% secure and unhackable. Yet "unhackable" is clearly what we need future AI systems to be before we put them in charge of, say, critical infrastructure or weapons systems, so the growing role of AI in society keeps raising the stakes for computer security. While some hacks exploit human gullibility or complex vulnerabilities in newly released software, others enable unauthorized login to remote computers by taking advantage of simple bugs that lingered unnoticed for an embarrassingly long time. The "Heartbleed" bug lasted from 2012 to 2014 in one of the most popular software libraries for secure communication between computers, and the "Bashdoor" bug was built into the very operating system of Unix computers from 1989 until 2014. This means that AI tools for improved verification and validation will improve security as well.

Unfortunately, better AI systems can also be used to find new vulnerabilities and perform more sophisticated hacks. Imagine, for example, that you one day get an unusually personalized "phishing" email attempting to persuade you to divulge personal information. It's sent from your friend's account by an AI who's hacked it and is impersonating her, imitating her writing style based on an analysis of her other sent emails, and including lots of personal information about you from other sources. Might you fall for this? What if the phishing email appears to come from your credit card company and is followed up by a phone call from a friendly human voice that you can't tell is AI-generated? In the ongoing computer-security arms race between offense and defense, there's so far little indication that defense is winning.

Laws

We humans are social animals who subdued all other species and conquered Earth thanks to our ability to cooperate. We've developed laws to incentivize and facilitate cooperation, so if AI can improve our legal and governance systems, then it can enable us to cooperate more successfully than ever before, bringing out the very best in us. And there's plenty of opportunity for improvement here, both in how our laws are applied and how they're written, so let's explore both in turn.

What are the first associations that come to your mind when you think about the court system in your country? If it's lengthy delays, high costs and occasional injustice, then you're not alone. Wouldn't it be wonderful if your first thoughts were instead "efficiency" and "fairness"? Since the legal process can be abstractly viewed as a computation, inputting information about evidence and laws and outputting a decision, some scholars dream of fully automating it with *robojudges:* AI systems that tirelessly apply the same high legal standards to every judgment without succumbing to human errors such as bias, fatigue or lack of the latest knowledge.

Robojudges

Byron De La Beckwith Jr. was convicted in 1994 of assassinating civil rights leader Medgar Evers in 1963, but two separate all-white Mississippi juries had failed to convict him the year after the murder, even though the physical evidence was essentially the same.[34] Alas, legal history is rife with judgments biased by skin color, gender, sexual orientation, religion, nationality and other factors. Robojudges could in principle ensure that, for the first time in history, everyone becomes truly equal under the law: they could be programmed to all be identical and to treat everyone equally, transparently applying the law in a truly unbiased fashion.

Robojudges could also eliminate human biases that are accidental rather than intentional. For example, a controversial 2012 study of Israeli judges claimed that they delivered significantly harsher verdicts when they were hungry: whereas they denied about 35% of parole cases right after breakfast, they denied over 85% right before lunch.[35] Another shortcoming of human judges is that they may lack sufficient time to explore all details of a case. In contrast, robojudges can easily be copied, since they consist of little more than software, allowing all pending cases to be processed in parallel rather than in series, each case getting its own robojudge for as long as it takes. Finally, although it's impossible for human judges to master all technical knowledge required for every possible case, from thorny patent disputes to murder mysteries hinging on the latest forensic science, future robojudges may have essentially unlimited memory and learning capacity.

One day, such robojudges may therefore be both more efficient and fairer, by virtue of being unbiased, competent and transparent. Their efficiency makes them fairer still: by speeding up the legal process and making it harder for savvy lawyers to skew the outcome, they could make it dramatically cheaper to get justice through the courts. This could greatly increase the chances of a cash-strapped individual or startup company prevailing against a billionaire or multinational corporation with an army of lawyers.

On the other hand, what if robojudges have bugs or get hacked? Both have already afflicted automatic voting machines, and when years behind bars or millions in the bank are at stake, the incentives for cyberattacks are greater still. Even if AI can be made robust enough for us to trust that a robojudge is using the legislated algorithm, will everybody feel that they understand its logical reasoning enough to respect its judgment? This challenge is exacerbated by the recent success of neural networks, which often outperform traditional easy-to-understand AI algorithms at the price of inscrutability. If defendants wish to know *why* they were convicted, shouldn't they have the right to a better answer than "we trained the system on lots of data, and this is what it decided"? Moreover, recent studies have shown that if you train a deep neural learning system with massive amounts of prisoner data, it can predict who's likely to return to crime (and should therefore be denied parole) better than human judges. But what if this system finds that recidivism is statistically linked to a prisoner's sex or race—would this count as a sexist, racist robojudge that needs reprogramming? Indeed, a 2016 study argued that recidivism-prediction software used across the United States was biased against African Americans and had contributed to unfair sentencing.[36] These are important questions that we all need to ponder and discuss to ensure that AI remains beneficial. We aren't facing an all-or-nothing decision regarding robojudges, but rather a decision about the extent and speed with which we want to deploy AI in our legal system. Do we want human judges to have AI-based decision support systems, just like tomorrow's medical doctors? Do we want to go further and have robojudge decisions that can be appealed to human judges, or do we want to go all the way and give even the final say to machines, even for death penalties?

Legal Controversies

So far, we've explored only the *application* of law; let us now turn to its *content*. There's broad consensus that our laws need to evolve to keep pace with our technology. For example, the two programmers who created the aforementioned ILOVEYOU worm and caused billions of dollars in damages were acquitted of all charges and walked free because at that time, there were no laws against malware creation in the Philippines. Since the pace of technological progress appears to be accelerating, laws need to be updated ever more rapidly, and have a tendency to lag behind. Getting more tech-savvy people into law schools and governments is probably a smart move for society. But should AI-based decision support systems for voters and legislators ensue, followed by outright robo-legislators?

How to best alter our laws to reflect AI progress is a fascinatingly controversial topic. One dispute reflects the tension between privacy versus freedom of information. Freedom fans argue that the less privacy we have, the more evidence the courts will have, and the fairer the judgments will be. For example, if the government taps into everyone's electronic devices to record where they are and what they type, click, say and do, many crimes would be readily solved, and additional ones could be prevented. Privacy advocates counter that they don't want an Orwellian surveillance state, and that even if they did, there's a risk of it turning into a totalitarian dictatorship of epic proportions. Moreover, machine-learning techniques have gotten better at analyzing brain data from fMRI scanners to determine what a person is thinking about and, in particular, whether they're telling the truth or lying.[37] If AI-assisted brain scanning technology became commonplace in courtrooms, the currently tedious process of establishing the facts of a case could be dramatically simplified and expedited, enabling faster trials and fairer judgments. But privacy advocates might worry about whether such systems occasionally make mistakes and, more fundamentally, whether our minds should be off-limits to government snooping. Governments that don't support freedom of thought could use such technology to criminalize the holding of certain beliefs and opinions. Where would *you* draw the line between justice and privacy, and between protecting society and protecting personal freedom? Wherever you draw it, will it gradually but inexorably move toward reduced privacy to compensate for the fact that evidence gets easier to fake? For example, once AI becomes able to generate fully realistic fake videos of you committing crimes, will you vote for a system where the government tracks everyone's whereabouts at all times and can provide you with an ironclad alibi if needed?

Another captivating controversy is whether AI research should be regulated or, more generally, what incentives policymakers should give AI researchers to maximize the chances of a beneficial outcome. Some AI researchers have argued against all forms of regulation of AI development, claiming that they would needlessly delay urgently needed innovation (for example, lifesaving self-driving cars) and would drive cutting-edge AI research underground and/or to other countries with more permissive governments. At the Puerto Rico beneficial-AI conference mentioned in the first chapter, Elon Musk argued that what we need right now from governments isn't oversight but insight: specifically, technically capable people in government positions who can monitor AI's progress and steer it if warranted down the road. He also argued that government regulation can sometimes nurture rather than stifle progress: for example, if government safety standards for self-driving cars can help reduce the number of self-driving-car accidents, then a public backlash is less likely and adoption of the new technology can be accelerated. The most safety-conscious AI companies might therefore favor regulation that forces less scrupulous competitors to match their high safety standards.

Yet another interesting legal controversy involves granting rights to machines. If self-driving cars cut the 32,000 annual U.S. traffic fatalities in half, perhaps carmakers won't get 16,000 thank-you notes, but 16,000 lawsuits. So if a self-driving car causes an accident, who should be liable—its occupants, its owner or its manufacturer? Legal scholar David Vladeck has proposed a fourth answer: the car itself! Specifically, he proposes that self-driving cars be allowed (and required) to hold car insurance. This way, models with a sterling safety record will qualify for premiums that are very low, probably lower than what's available to human drivers, while poorly designed models from sloppy manufacturers will only qualify for insurance policies that make them prohibitively expensive to own.

But if machines such as cars are allowed to hold insurance policies, should they also be able to own money and property? If so, there's nothing legally stopping smart computers from making money on the stock market and using it to buy online services. Once a computer starts paying humans to work for it, it can accomplish anything that humans can do. If AI systems eventually get better than humans at investing (which they already are in some domains), this could lead to a situation where most of our economy is owned and controlled by machines. Is this what we want? If it sounds far-off, consider that most of our economy is already owned by another form of non-human entity: corporations, which are often more powerful than any one person in them and can to some extent take on life of their own.

If you're OK with granting machines the rights to own property, then how about granting them the right to vote? If so, should each computer program get one vote, even though it can trivially make trillions of copies of itself in the cloud if it's rich enough, thereby guaranteeing that it will decide all elections? If not, then on what moral basis are we discriminating against machine minds relative to human minds? Does it make a difference if machine minds are conscious in the sense of having a subjective experience like we do? We'll explore in greater depth these controversial questions related to computer control of our world in the next chapter, and questions related to machine consciousness in chapter 8.

Weapons

Since time immemorial, humanity has suffered from famine, disease and war. We've already mentioned how AI may help reduce famine and disease, so how about war? Some argue that nuclear weapons deter war between the countries that own them because they're so horrifying, so how about letting all nations build even more horrifying AI-based weapons in the hope of ending all war forever? If you're unpersuaded by that argument and believe that future wars are inevitable, how about using AI to make these wars more humane? If wars consist merely of machines fighting machines, then no human soldiers or civilians need get killed. Moreover, future AI-powered drones and other autonomous weapon systems (AWS; also known by their opponents as "killer robots") can hopefully be made more fair and rational than human soldiers: equipped with superhuman sensors and unafraid of getting killed, they might remain cool, calculating and level-headed even in the heat of battle, and be less likely to accidentally kill civilians.

Figure 3.4: Whereas today's military drones (such as this U.S. Air Force MQ-1 Predator) are remote-controlled by humans, future AI-powered drones have the potential to take humans out of the loop, using an algorithm to decide whom to target and kill.

A Human in the Loop

But what if the automated systems are buggy, confusing or don't behave as expected? The U.S. Phalanx system for Aegis-class cruisers automatically detects, tracks and attacks threats such as anti-ship missiles and aircraft. The USS *Vincennes* was a guided missile cruiser nicknamed Robocruiser in reference to its Aegis system, and on July 3, 1988, in the midst of a skirmish with Iranian gunboats during the Iran-Iraq war, its radar system warned of an incoming aircraft. Captain William Rodgers III inferred that they were being attacked by a diving Iranian F-14 fighter jet and gave the Aegis system approval to fire. What he didn't realize at the time was that they shot down Iran Air Flight 655, a civilian Iranian passenger jet, killing all 290 people on board and causing international outrage. Subsequent investigation implicated a confusing user interface that didn't automatically show which dots on the radar screen were civilian planes (Flight 655 followed its regular daily flight path and had its civilian aircraft transponder on) or which dots were descending (as for an attack) vs. ascending (as Flight 655 was doing after takeoff from Tehran). Instead, when the automated system was queried for information about the mysterious aircraft, it reported "descending" because that was the status of a different aircraft to which it had confusingly reassigned a number used by the navy to track planes: what was descending was instead a U.S. surface combat air patrol plane operating far away in the Gulf of Oman.

In this example, there was a human in the loop making the final decision, who under time pressure placed too much trust in what the automated system told him. So far, according to defense officials around the world, all deployed weapons systems have a human in the loop, with the exception of low-tech booby traps such as land mines. But development is now under way of truly autonomous weapons that select and attack targets entirely on their own. It's militarily tempting to take all humans out of the loop to gain speed: in a dogfight between a fully autonomous drone that can respond instantly and a drone reacting more sluggishly because it's remote-controlled by a human halfway around the world, which one do you think would win?

However, there have been close calls where we were extremely lucky that there was a human in the loop. On October 27, 1962, during the Cuban Missile Crisis, eleven U.S. Navy destroyers and the aircraft carrier USS *Randolph* had cornered the Soviet submarine B-59 near Cuba, in international waters outside the U.S. "quarantine" area. What they didn't know was that the temperature onboard had risen past 45°C (113°F) because the submarine's batteries were running out and the air-conditioning had stopped. On the verge of carbon dioxide poisoning, many crew members had fainted. The crew had had no contact with Moscow for days and didn't know whether World War III had already begun. Then the Americans started dropping small depth charges, which they had, unbeknownst to the crew, told Moscow were merely meant to force the sub to surface and leave. "We thought—that's it—the end," crew member V. P. Orlov recalled. "It felt like you were sitting in a metal barrel, which somebody is constantly blasting with a sledgehammer." What the Americans also didn't know was that the B-59 crew had a nuclear torpedo that they were authorized to launch without clearing it with Moscow. Indeed, Captain Savitski decided to launch the nuclear torpedo. Valentin Grigorievich, the torpedo officer, exclaimed: "We will die, but we will sink them all—we will not disgrace our navy!" Fortunately, the decision to launch had to be authorized by three officers on board, and one of them, Vasili Arkhipov, said no. It's sobering that very few have heard of Arkhipov, although his decision may have averted World War III and been the single most valuable contribution to humanity in modern history.[38] It's also sobering to contemplate what might have happened had B-59 been an autonomous AI-controlled submarine with no humans in the loop.

Two decades later, on September 9, 1983, tensions were again high between the superpowers: the Soviet Union had recently been called an "evil empire" by U.S. president Ronald Reagan, and just the previous week, it had shot down a Korean Airlines passenger plane that strayed into its airspace, killing 269 people—including a U.S. congressman. Now an automated Soviet early-warning system reported that the United States had launched five land-based nuclear missiles at the Soviet Union, leaving Officer Stanislav Petrov merely minutes to decide whether this was a false alarm. The satellite was found to be operating properly, so following protocol would have led him to report an incoming nuclear attack. Instead, he trusted his gut instinct, figuring that the United States was unlikely to attack with only five missiles, and reported to his commanders that it was a false alarm without knowing this to be true. It later became clear that a satellite had mistaken the Sun's reflections off cloud tops for flames from rocket engines.[39] I wonder what would have happened if Petrov had been replaced by an AI system that properly followed proper protocol.

The Next Arms Race?

As you've undoubtedly guessed by now, I personally have serious concerns about autonomous weapons systems. But I haven't even begun to tell you about my main worry: the endpoint of an arms race in AI weapons. In July 2015, I expressed this worry in the following open letter together with Stuart Russell, with helpful feedback from my colleagues at the Future of Life Institute:[40]

AUTONOMOUS WEAPONS:

An Open Letter from AI & Robotics Researchers

Autonomous weapons select and engage targets without human intervention. They might include, for example, armed quadcopters that can search for and eliminate people meeting certain predefined criteria, but do not include cruise missiles or remotely piloted drones for which humans make all targeting decisions. Artificial Intelligence (AI) technology has reached a point where the deployment of such systems is practically if not legally feasible within years, not decades, and the stakes are high: autonomous weapons have been described as the third revolution in warfare, after gunpowder and nuclear arms.

Many arguments have been made for and against autonomous weapons, for example that replacing human soldiers by machines is good by reducing casualties for the owner but bad by thereby lowering the threshold for going to battle. The key question for

> humanity today is whether to start a global AI arms race or to prevent it from starting. If any major military power pushes ahead with AI weapon development, a global arms race is virtually inevitable, and the endpoint of this technological trajectory is obvious: autonomous weapons will become the Kalashnikovs of tomorrow. Unlike nuclear weapons, they require no costly or hard-to-obtain raw materials, so they'll become ubiquitous and cheap for all significant military powers to mass-produce. It will only be a matter of time until they appear on the black market and in the hands of terrorists, dictators wishing to better control their populace, warlords wishing to perpetrate ethnic cleansing, etc. Autonomous weapons are ideal for tasks such as assassinations, destabilizing nations, subduing populations and selectively killing a particular ethnic group. We therefore believe that a military AI arms race would not be beneficial for humanity. There are many ways in which AI can make battlefields safer for humans, especially civilians, without creating new tools for killing people.
>
> Just as most chemists and biologists have no interest in building chemical or biological weapons, most AI researchers have no interest in building AI weapons and do not want others to tarnish their field by doing so, potentially creating a major public backlash against AI that curtails its future societal benefits. Indeed, chemists and biologists have broadly supported international agreements that have successfully prohibited chemical and biological weapons, just as most physicists supported the treaties banning space-based nuclear weapons and blinding laser weapons.

To make it harder to dismiss our concerns as coming only from pacifist tree-huggers, I wanted to get our letter signed by as many hardcore AI researchers and roboticists as possible. The International Campaign for Robotic Arms Control had previously amassed hundreds of signatories who called for a ban on killer robots, and I suspected that we could do even better. I knew that professional organizations would be reluctant to share their massive member email lists for a purpose that could be construed as political, so I scraped together lists of researchers' names and institutions from online documents and advertised the task of finding their email addresses on MTurk—the Amazon Mechanical Turk crowdsourcing platform. Most researchers have their email addresses listed on their university websites, and twenty-four hours and $54 later, I was the proud owner of a mailing list of hundreds of AI researchers who'd been successful enough to be elected Fellows of the Association for the Advancement of Artificial Intelligence (AAAI). One of them was the British-Australian AI professor Toby Walsh, who kindly agreed to email everyone else on the list and help spearhead our campaign. MTurk workers around the world tirelessly produced additional mailing lists for Toby, and before long, over 3,000 AI researchers and robotics researchers had signed our open letter, including six past AAAI presidents and AI industry leaders from Google, Facebook, Microsoft and Tesla. An army of FLI volunteers worked tirelessly to validate the signatory lists, removing spoof entries such as Bill Clinton and Sarah Connor. Over 17,000 others signed too, including Stephen Hawking, and after Toby organized a press conference about this at the International Joint Conference of Artificial Intelligence, it became a major news story around the world.

Because biologists and chemists once took a stand, their fields are now known mainly for creating beneficial medicines and materials rather than biological and chemical weapons. The AI and robotics communities had now spoken as well: the letter signatories also wanted their fields to be known for creating a better future, not for creating new ways of killing people. But will the main future use of AI be civilian or military? Although we've spent more pages in this chapter on the former, we may soon be spending more money on the latter—especially if a military AI arms race takes off. Civilian AI investment commitments exceeded a billion dollars in 2016, but this was dwarfed by the Pentagon's fiscal 2017 budget request of $12–15 billion for AI-related projects, and China and Russia are likely to take note of what Deputy Defense Secretary Robert Work said when this was announced: "I want our competitors to wonder what's behind the black curtain."[41]

Should There Be an International Treaty?

Although there's now a major international push toward negotiating some form of killer robot ban, it's still unclear what will happen, and there's a vibrant ongoing debate about what, if anything, *should* happen. Although many leading stakeholders agree that world powers should draft some form of international regulations to guide AWS research and use, there's less agreement about what precisely should be banned and how a ban would be enforced. For example, should only lethal autonomous weapons be banned, or also ones that seriously injure people, say by blinding them? Would we ban development, production or ownership? Should a ban apply to all autonomous weapons systems or, as our letter said, only offensive ones, allowing defensive systems such as autonomous anti-aircraft guns and missile defenses? In the latter case, should AWS count as defensive even if they're easy to move into enemy territory? And how would you enforce a treaty given that most components of an autonomous weapon have a dual civilian use as well? For example, there isn't much difference between a drone that can deliver Amazon packages and one that can deliver bombs.

Some debaters have argued that designing an effective AWS treaty is hopelessly hard and that we therefore shouldn't even try. On the other hand, John F. Kennedy emphasized when announcing the Moon missions that hard things are worth attempting when success will greatly benefit the future of humanity. Moreover, many experts argue that the bans on biological and chemical weapons were valuable even though enforcement proved hard, with significant cheating, because the bans caused severe stigmatization that limited their use.

I met Henry Kissinger at a dinner event in 2016, and got the opportunity to ask him about his role in the biological weapons ban. He explained how back when he was the U.S. national security adviser, he'd persuaded President Nixon that a ban would be good for U.S. national security. I was impressed by how sharp his mind and memory were for a ninety-two-year-old, and was fascinated to hear his inside perspective. Since the United States already enjoyed superpower status thanks to its conventional and nuclear forces, it had more to lose than to gain from a worldwide bioweapons arms race with uncertain outcome. In other words, if you're already top dog, then it makes sense to follow the maxim "If it ain't broke, don't fix it." Stuart Russell joined our after-dinner conversation, and we discussed how exactly the same argument can be made about lethal autonomous weapons: those who stand to gain most from an arms race aren't superpowers but small rogue states and non-state actors such as terrorists, who gain access to the weapons via the black market once they've been developed.

Once mass-produced, small AI-powered killer drones are likely to cost little more than a smartphone. Whether it's a terrorist wanting to assassinate a politician or a jilted lover seeking revenge on his ex-girlfriend, all they need to do is upload their target's photo and address into the killer drone: it can then fly to the destination, identify and eliminate the person, and self-destruct to ensure that nobody knows who was responsible. Alternatively, for those bent on ethnic cleansing, it can easily be programmed to kill only people with a certain skin color or ethnicity. Stuart envisions that the smarter such weapons get, the less material, firepower and money will be needed per kill. For example, he fears bumblebee-sized drones that kill cheaply using minimal explosive power by shooting people in the eye, which is soft enough to allow even a small projectile to continue into the brain. Or they might latch on to the head with metal claws and then penetrate the skull with a tiny shaped charge. If a million such killer drones can be dispatched from the back of a single truck, then one has a horrifying weapon of mass destruction of a whole new kind: one that can selectively kill only a prescribed category of people, leaving everybody and everything else unscathed.

A common counterargument is that we can eliminate such concerns by making killer robots ethical—for example, so that they'll only kill enemy soldiers. But if we worry about enforcing a ban, then how would it be easier to enforce a requirement that enemy autonomous weapons be 100% ethical than to enforce that they aren't produced in the first place? And can one consistently claim that the well-trained soldiers of civilized nations are so bad at following the rules of war that robots can do better, while at the same time claiming that rogue nations, dictators and terrorist groups are so good at following the rules of war that they'll never choose to deploy robots in ways that violate these rules?

Cyberwar

Another interesting military aspect of AI is that it may let you attack your enemy even without building any weapons of your own, through cyberwarfare. As a small prelude to what the future may bring, the Stuxnet worm, widely attributed to the U.S. and Israeli governments, infected fast-spinning centrifuges in Iran's nuclear-enrichment program and caused them to tear themselves apart. The more automated society gets and the more powerful the attacking AI becomes, the more devastating cyberwarfare can be. If you can hack and crash your enemy's self-driving cars, auto-piloted planes, nuclear reactors, industrial robots, communication systems, financial systems and power grids, then you can effectively crash his economy and cripple his defenses. If you can hack some of his weapons systems as well, even better.

We began this chapter by surveying how spectacular the near-term opportunities are for AI to benefit humanity—if we manage to make it robust and unhackable. Although AI itself can be used to make AI systems more robust, thereby aiding the cyberwar defense, AI can clearly aid the offense as well. Ensuring that the defense prevails must be one of the most crucial short-term goals for AI development—otherwise all the awesome technology we build can be turned against us!

Jobs and Wages

So far in this chapter, we've mainly focused on how AI will affect us as *consumers*, by enabling transformative new products and services at affordable prices. But how will it affect us as *workers*, by transforming the job market? If we can figure out how to grow our prosperity through automation without leaving people lacking income or purpose, then we have the potential to create a fantastic future with leisure and unprecedented opulence for everyone who wants it. Few people have thought longer and harder about this than economist Erik Brynjolfsson, one of my MIT colleagues. Although he's always well-groomed and impeccably dressed, he has Icelandic heritage, and I sometimes can't help imagine that he only recently trimmed back a wild red Viking beard and mane to blend in at our business school. He certainly hasn't trimmed back his wild ideas, and he calls his optimistic job-market vision "Digital Athens." The reason that the Athenian citizens of antiquity had lives of leisure where they could enjoy democracy, art and games was mainly that they had slaves to do much of the work. But why not replace the slaves with AI-powered robots, creating a digital utopia that everyone can enjoy? Erik's AI-driven economy would not only eliminate stress and drudgery and produce an abundance of everything we want today, but it would also supply a bounty of wonderful new products and services that today's consumers haven't yet realized that they want.

Technology and Inequality

We can get from where we are today to Erik's Digital Athens if everyone's hourly salary keeps growing year by year, so that those who want more leisure can gradually work less while continuing to improve their standard of living. Figure 3.5 shows that this is precisely what happened in the United States from World War II until the mid-1970s: although there was income inequality, the total size of the pie grew in such a way that almost everybody got a larger slice. But then, as Erik is the first to admit, something changed: figure 3.5 shows that although the economy kept growing and raising the average income, the gains over the past four decades went to the wealthiest, mostly to the top 1%, while the poorest 90% saw their incomes stagnate. The resulting growth in inequality is even more evident if we look not at income but at wealth. For the bottom 90% of U.S. households, the average net worth was about $85,000 in 2012—the same as twenty-five years earlier—while the top 1% more than doubled their inflation-adjusted wealth during that period, to $14 million.[42] Differences are even more extreme internationally where, in 2013, the combined wealth of the bottom half of the world's population (over 3.6 billion people) is the same as that of the world's eight richest people[43]—a statistic that highlights the poverty and vulnerability at the bottom as much as the wealth at the top. At our 2015 Puerto Rico conference, Erik told the assembled AI researchers that he thought progress in AI and automation would continue making the economic pie bigger, but that there's no economic law that everyone, or even most people, will benefit.

Although there's broad agreement among economists that inequality is rising, there's an interesting controversy about why and whether the trend will continue. Debaters on the left side of the political spectrum often argue that the main cause is globalization and/or economic policies such as tax cuts for the rich. But Erik Brynjolfsson and his MIT collaborator Andrew McAfee argue that the main cause is something else: technology.[44] Specifically, they argue that digital technology drives inequality in three different ways.

First, by replacing old jobs with ones requiring more skills, technology has rewarded the educated: since the mid-1970s, salaries rose about 25% for those with graduate degrees while the average high school dropout took a 30% pay cut.[45]

Second, they claim that since the year 2000, an ever-larger share of corporate income has gone to those who own the companies as opposed to those who work there—and that as long as automation continues, we should expect those who own the machines to take a growing fraction of the pie. This edge of capital over labor may be particularly important for the growing digital economy, which tech visionary Nicholas Negroponte defines as moving bits, not atoms.

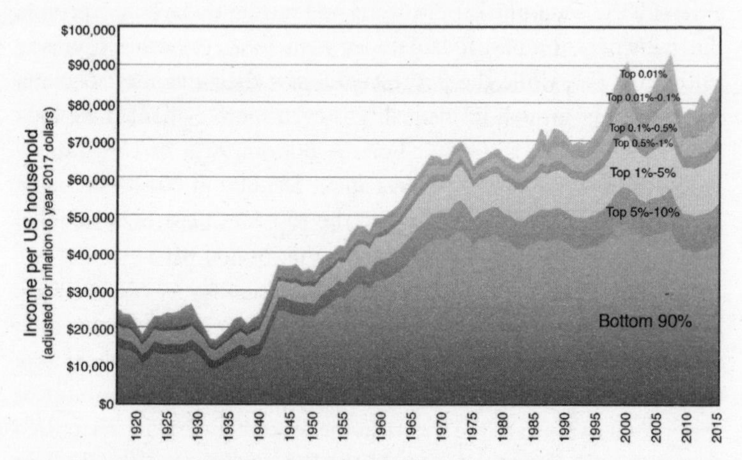

Figure 3.5: How the economy has grown average income over the past century, and what fraction of this income has gone to different groups. Before the 1970s, rich and poor are seen to all be getting better off in lockstep, after which most of the gains have gone to the top 1% while the bottom 90% have on average gained close to nothing.[46] The amounts have been inflation-corrected to year-2017 dollars.

Now that everything from books to movies and tax preparation tools has gone digital, additional copies can be sold worldwide at essentially zero cost, without hiring additional employees. This allows most of the revenue to go to investors rather than workers, and helps explain why, even though the combined revenues of Detroit's "Big 3" (GM, Ford and Chrysler) in 1990 were almost identical to those of Silicon Valley's "Big 3" (Google, Apple, Facebook) in 2014, the latter had nine times fewer employees and were worth thirty times more on the stock market.[47]

Third, Erik and collaborators argue that the digital economy often benefits superstars over everyone else. Harry Potter author J. K. Rowling became the first writer to join the billionaire club, and she got much richer than Shakespeare because her stories could be transmitted in the form of text, movies and games to billions of people at very low cost. Similarly, Scott Cook made a billion on the TurboTax tax preparation software, which, unlike human tax preparers, can be sold as a download. Since most people are willing to pay little or nothing for the tenth-best tax-preparation software, there's room in the marketplace for only a modest number of superstars. This means that if all the world's parents advise their kids to become the next J. K. Rowling, Gisele Bündchen, Matt Damon, Cristiano Ronaldo, Oprah Winfrey or Elon Musk, almost none of their kids will find this a viable career strategy.

Career Advice for Kids

So what career advice *should* we give our kids? I'm encouraging mine to go into professions that machines are currently bad at, and therefore seem unlikely to get automated in the near future. Recent forecasts for when various jobs will get taken over by machines identify several useful questions to ask about a career before deciding to educate oneself for it.[48] For example:

- Does it require interacting with people and using social intelligence?
- Does it involve creativity and coming up with clever solutions?
- Does it require working in an unpredictable environment?

The more of these questions you can answer with a yes, the better your career choice is likely to be. This means that relatively safe bets include becoming a teacher, nurse, doctor, dentist, scientist, entrepreneur, programmer, engineer, lawyer, social worker, clergy member, artist, hairdresser or massage therapist.

In contrast, jobs that involve highly repetitive or structured actions in a predictable setting aren't likely to last long before getting automated away. Computers and industrial robots took over the simplest such jobs long ago, and improving technology is in the process of eliminating many more, from telemarketers to warehouse workers, cashiers, train operators, bakers and line cooks.[49] Drivers of trucks, buses, taxis and Uber/Lyft cars are likely to follow soon. There are many more professions (including paralegals, credit analysts, loan officers, bookkeepers and tax accountants) that, although they aren't on the endangered list for full extinction, are getting most of their tasks automated and therefore demand many fewer humans.

But staying clear of automation isn't the only career challenge. In this global digital age, aiming to become a professional writer, filmmaker, actor, athlete or fashion designer is risky for another reason: although people in these professions won't get serious competition from machines anytime soon, they'll get increasingly brutal competition from other humans around the globe according to the aforementioned superstar theory, and very few will succeed.

In many cases, it would be too myopic and crude to give career advice at the level of whole fields: there are many jobs that won't get entirely eliminated, but which will see many of their tasks automated. For example, if you go into medicine, don't be the radiologist who analyzes the medical images and gets replaced by IBM's Watson, but the doctor who orders the radiology analysis, discusses the results with the patient, and decides on the treatment plan. If you go into finance, don't be the "quant" who applies algorithms to the data and gets replaced by software, but the fund manager who uses the quantitative analysis results to make strategic investment decisions. If you go into law, don't be the paralegal who reviews thousands of documents for the discovery phase and gets automated away, but the attorney who counsels the client and presents the case in court.

So far, we've explored what individuals can do to maximize their success on the job market in the age of AI. But what can governments do to help their workforces succeed? For example, what education system best prepares people for a job market where AI keeps improving rapidly? Is it still our current model with one or two decades of education followed by four decades of specialized work? Or is it better to switch to a system where people work for a few years, then go back to school for a year, then work for a few more years?[50] Or should continuing education (perhaps provided online) be a standard part of any job?

And what economic policies are most helpful for creating good new jobs? Andrew McAfee argues that there are many policies that are likely to help, including investing heavily in research, education and infrastructure, facilitating migration and incentivizing entrepreneurship. He feels that "the Econ 101 playbook is clear, but is not being followed," at least not in the United States.[51]

Will Humans Eventually Become Unemployable?

If AI keeps improving, automating ever more jobs, what will happen? Many people are job optimists, arguing that the automated jobs will be replaced by new ones that are even better. After all, that's what's always happened before, ever since Luddites worried about technological unemployment during the Industrial Revolution.

Others, however, are job pessimists and argue that this time is different, and that an ever-larger number of people will become not only unemployed, but unemployable.[52] The job pessimists argue that the free market sets salaries based on supply and demand, and that a growing supply of cheap machine labor will eventually depress human salaries far below the cost of living. Since the market salary for a job is the hourly cost of whoever or whatever will perform it most cheaply, salaries have historically dropped whenever it became possible to outsource a particular occupation to a lower-income country or to a cheap machine. During the Industrial Revolution, we started figuring out how to replace our muscles with machines, and people shifted into better-paying jobs where they used their minds more. Blue-collar jobs were replaced by white-collar jobs. Now we're gradually figuring out how to replace our minds by machines. If we ultimately succeed in this, then what jobs are left for us?

Some job optimists argue that after physical and mental jobs, the next boom will be in *creative* jobs, but job pessimists counter that creativity is just another mental process, so that it too will eventually be mastered by AI. Other job optimists hope that the next boom will instead be in new technology-enabled professions that we haven't even thought of yet. After all, who during the Industrial Revolution would have imagined that their descendants would one day work as web designers and Uber drivers? But job pessimists counter that this is wishful thinking with little support from empirical data. They point out that we could have made the same argument a century ago, before the computer revolution, and predicted that most of today's professions would be new and previously unimagined technology-enabled ones that didn't use to exist. This prediction would have been an epic failure, as illustrated in figure 3.6: the vast majority of today's occupations are ones that already existed a century ago, and when we sort them by the number of jobs they provide, we have to go all the way down to twenty-first place in the list until we encounter a new occupation: software developers, who make up less than 1% of the U.S. job market.

We can get a better understanding of what's happening by recalling figure 2.2 from chapter 2, which showed the landscape of human intelligence, with elevation representing how hard it is for machines to perform various tasks and the rising sea level showing what machines can currently do. The main trend on the job market isn't that we're moving into entirely new professions. Rather, we're crowding into those pieces of terrain in figure 2.2 that haven't yet been submerged by the rising tide of technology! Figure 3.6 shows that this forms not a single island but a complex archipelago, with islets and atolls corresponding to all the valuable things that machines still can't do as cheaply as humans can. This includes not only high-tech professions such as software development, but also a panoply of low-tech jobs leveraging our superior dexterity and social skills, ranging from massage therapy to acting. Might AI eclipse us at intellectual tasks so rapidly that the last remaining jobs will be in that low-tech category?

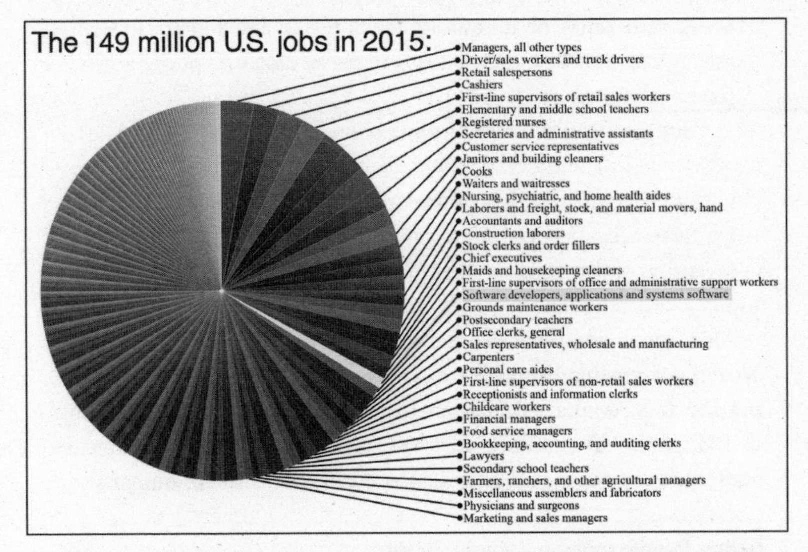

Figure 3.6: The pie chart shows the occupations of the 149 million Americans who had a job in 2015, with the 535 job categories from the U.S. Bureau of Labor Statistics sorted by popularity.[53] All occupations with more than a million workers are labeled. There are no new occupations created by computer technology until twenty-first place. This figure is inspired by an analysis from Federico Pistono.[54]

A friend of mine recently joked with me that perhaps the very last profession will be the very first profession: prostitution. But then he mentioned this to a Japanese roboticist, who protested: "No, robots are very good at those things!"

Job pessimists contend that the endpoint is obvious: the whole archipelago will get submerged, and there will be no jobs left that humans can do more cheaply than machines. In his 2007 book *Farewell to Alms*, the Scottish-American economist Gregory Clark points out that we can learn a thing or two about our future job prospects by comparing notes with our equine friends. Imagine two horses looking at an early automobile in the year 1900 and pondering their future.

"I'm worried about technological unemployment."

"Neigh, neigh, don't be a Luddite: our ancestors said the same thing when steam engines took our industry jobs and trains took our jobs pulling stage coaches. But we have more jobs than ever today, and they're better too: I'd much rather pull a light carriage through town than spend all day walking in circles to power a stupid mine-shaft pump."

"But what if this internal combustion engine thing really takes off?"

"I'm sure there'll be new new jobs for horses that we haven't yet imagined. That's what's always happened before, like with the invention of the wheel and the plow."

Alas, those not-yet-imagined new jobs for horses never arrived. No-longer-needed horses were slaughtered and not replaced, causing the U.S. equine population to collapse from about 26 million in 1915 to about 3 million in 1960.[55] As mechanical muscles made horses redundant, will mechanical minds do the same to humans?

Giving People Income Without Jobs

So who's right: those who say automated jobs will be replaced by better ones or those who say most humans will end up unemployable? If AI progress continues unabated, then *both* sides might be right: one in the short term and the other in the long term. But although people often discuss the disappearance of jobs with doom-and-gloom connotations, it doesn't have to be a bad thing! Luddites obsessed about particular jobs, neglecting the possibility that other jobs might provide the same social value. Analogously, perhaps those who obsess about jobs today are being too narrow-minded: we want jobs because they can provide us with income and purpose, but given the opulence of resources produced by machines, it should be possible to find alternative ways of providing both the income and the purpose *without* jobs. Something similar ended up happening in the equine story, which didn't end with all horses going extinct. Instead, the number of horses has more than tripled since 1960, as they were protected by an equine social-welfare system of sorts: even though they couldn't pay their own bills, people decided to take care of horses, keeping them around for fun, sport and companionship. Can we similarly take care of our fellow humans in need?

Let's start with the question of income: redistributing merely a small share of the growing economic pie should enable everyone to become better off. Many argue that we not only *can* but *should* do this. On the 2016 panel where Moshe Vardi spoke of a moral imperative to save lives with AI-powered technology, I argued that it's also a moral imperative to advocate for its beneficial use, including sharing the wealth. Erik Brynjolfsson, also a panelist, said that "if with all this new wealth generation, we can't even prevent half of all people from getting worse off, then shame on us!"

There are many different proposals for wealth-sharing, each with its supporters and detractors. The simplest is *basic income*, where every person receives a monthly payment with no preconditions or requirements whatsoever. A number of small-scale experiments are now being tried or planned, for example in Canada, Finland and the Netherlands. Advocates argue that basic income is more efficient than alternatives such as welfare payments to the needy, because it eliminates the administrative hassle of determining who qualifies. Need-based welfare payments have also been criticized for disincentivizing work, but this of course becomes irrelevant in a jobless future where nobody works.

Governments can help their citizens not only by giving them money, but also by providing them with free or subsidized services such as roads, bridges, parks, public transportation, childcare, education, healthcare, retirement homes and internet access; indeed, many governments already provide most of these services. As opposed to basic income, such government-funded services accomplish two separate goals: they reduce people's cost of living and also provide jobs. Even in a future where machines can outperform humans at all jobs, governments could opt to pay people to work in childcare, eldercare, etc. rather than outsource the caregiving to robots.

Interestingly, technological progress can end up providing many valuable products and services for free even without government intervention. For example, people used to pay for encyclopedias, atlases, sending letters and making phone calls, but now anyone with an internet connection gets access to all these things at no cost—together with free videoconferencing, photo sharing, social media, online courses and countless other new services. Many other things that can be highly valuable to a person, say a lifesaving course of antibiotics, have become extremely cheap. So thanks to technology, even many poor people today have access to things that the world's richest people lacked in the past. Some take this to mean that the income needed for a decent life is dropping.

If machines can one day produce all current goods and services at minimal cost, then there's clearly enough wealth to make everyone better off. In other words, even relatively modest taxes could then allow governments to pay for basic income and free services. But the fact that wealth-sharing *can* happen obviously doesn't mean that it *will* happen, and today there's strong political disagreement about whether it even *should* happen. As we saw above, the current trend in the United States appears to be in the opposite direction, with some groups of people getting poorer decade after decade. Policy decisions about how to share society's growing wealth will impact everybody, so the conversation about what sort of future economy to build should include everyone, not merely AI researchers, roboticists and economists.

Many debaters argue that reducing income inequality is a good idea not merely in an AI-dominated future, but also today. Although the main argument tends to be a moral one, there's also evidence that greater equality makes democracy work better: when there's a large well-educated middle class, the electorate is harder to manipulate, and it's tougher for small numbers of people or companies to buy undue influence over the government. A better democracy can in turn enable a better-managed economy that's less corrupt, more efficient and faster growing, ultimately benefiting essentially everyone.

Giving People Purpose Without Jobs

Jobs can provide people with more than just money. Voltaire wrote in 1759 that "work keeps at bay three great evils: boredom, vice and need." Conversely, providing people with income isn't enough to guarantee their well-being. Roman emperors provided both bread and circuses to keep their underlings content, and Jesus emphasized non-material needs in the Bible quote "Man shall not live by bread alone." So precisely what valuable things do jobs contribute beyond money, and in what alternative ways can a jobless society provide them?

The answers to these questions are obviously complicated, since some people hate their jobs and others love them. Moreover, many children, students and homemakers thrive without jobs, while history teems with stories of spoiled heirs and princes who succumbed to ennui and depression. A 2012 meta-analysis showed that unemployment tends to have negative long-term effects on well-being, while retirement was a mixed bag with both positive and negative aspects.[56] The growing field of *positive psychology* has identified a number of factors that boost people's sense of well-being and purpose, and found that some (but not all!) jobs can provide many of them, for example:[57]

- a social network of friends and colleagues
- a healthy and virtuous lifestyle
- respect, self-esteem, self-efficacy and a pleasurable sense of "flow" stemming from doing something one is good at
- a sense of being needed and making a difference
- a sense of meaning from being part of and serving something larger than oneself

This gives reason for optimism, since all of these things can be provided also outside of the workplace, for example through sports, hobbies and learning, and with families, friends, teams, clubs, community groups, schools, religious and humanist organizations, political movements and other institutions. To create a low-employment society that flourishes rather than degenerates into self-destructive behavior, we therefore need to understand how to help such well-being-inducing activities thrive. The quest for such an understanding needs to involve not only scientists and economists, but also psychologists, sociologists and educators. If serious efforts are put into creating well-being for all, funded by part of the wealth that future AI generates, then society should be able to flourish like never before. At a minimum, it should be possible to make everyone as happy as if they had their personal dream job, but once one breaks free of the constraint that everyone's activities must generate income, the sky's the limit.

Human-Level Intelligence?

We've explored in this chapter how AI has the potential to greatly improve our lives in the near term, as long as we plan ahead and avoid various pitfalls. But what about the longer term? Will AI progress eventually stagnate due to insurmountable obstacles, or will AI researchers ultimately succeed in their original goal of building human-level artificial general intelligence? We saw in the previous chapter how the laws of physics allow suitable clumps of matter to remember, compute and learn, and how they don't prohibit such clumps from one day doing so with greater intelligence than the matter clumps in our heads. If/when we humans will succeed in building such superhuman AGI is much less clear. We saw in the first chapter that we simply don't know yet, since the world's leading AI experts are divided, most of them making estimates ranging from decades to centuries and some even guessing never. Forecasting is tough because, when you're exploring uncharted territory, you don't know how many mountains separate you from your destination. Typically you see only the closest one, and need to climb it before you can discover your next obstacle.

What's the soonest it could happen? Even if we knew the best possible way to build human-level AGI using today's computer hardware, which we don't, we'd still need to have enough of it to provide the raw computational power needed. So what's the computational power of a human brain measured in the bits and FLOPS from chapter 2?* This is a delightfully tricky question, and the answer depends dramatically on how we ask it:

- Question 1: How many FLOPS are needed to simulate a brain?
- Question 2: How many FLOPS are needed for human intelligence?
- Question 3: How many FLOPS can a human brain perform?

There have been lots of papers published on question 1, and they typically give answers in the ballpark of a hundred petaFLOPS, i.e., 10^{17} FLOPS.[58] That's about the same computational power as the Sunway TaihuLight (figure 3.7), the world's fastest supercomputer in

* Recall that FLOPS are floating-point operations per second, say, how many 19-digit numbers can be multiplied each second.

2016, which cost about $300 million. Even if we knew how to use it to simulate the brain of a highly skilled worker, we would only profit from having the simulation do this person's job if we could rent the TaihuLight for less than her hourly salary. We may need to pay even more, because many scientists believe that to accurately replicate the intelligence of a brain, we can't treat it as a mathematically simplified neural-network model from chapter 2. Perhaps we instead need to simulate it at the level of individual molecules or even subatomic particles, which would require dramatically more FLOPS.

The answer to question 3 is easier: I'm painfully bad at multiplying 19-digit numbers, and it would take me many minutes even if you let me borrow pencil and paper. That would clock me in below 0.01 FLOPS—a whopping 19 orders of magnitude below the answer to question 1! The reason for the huge discrepancy is that brains and supercomputers are optimized for extremely different tasks. We get a similar discrepancy between these questions:

How well can a tractor do the work of a Formula One race car?
How well can a Formula One car do the work of a tractor?

So which of these two questions about FLOPS are we trying to answer to forecast the future of AI? Neither! If we wanted to simulate a human brain, we'd care about question 1, but to build human-level AGI, what matters is instead the one in the middle: question 2. Nobody knows its answer yet, but it may well be significantly cheaper than simulating a brain if we either adapt the software to be better matched to today's computers or build more brain-like hardware (rapid progress is being made on so-called neuromorphic chips).

Hans Moravec estimated the answer by making an apples-to-apples comparison for a computation that both our brain and today's computers can do efficiently: certain low-level image-processing tasks that a human retina performs in the back of the eyeball before sending its results to the brain via the optic nerve.[59] He figured that replicating a retina's computations on a conventional computer requires about a billion FLOPS and that the whole brain does about ten thousand times more computation than a retina (based on comparing volumes

Figure 3.7: Sunway TaihuLight, the world's fastest supercomputer in 2016, whose raw computational power arguably exceeds that of the human brain.

and numbers of neurons), so that the computational capacity of the brain is around 10^{13} FLOPS—roughly the power of an optimized $1,000 computer in 2015!

In summary, there's absolutely no guarantee that we'll manage to build human-level AGI in our lifetime—or ever. But there's also no watertight argument that we won't. There's no longer a strong argument that we lack enough hardware firepower or that it will be too expensive. We don't know how far we are from the finish line in terms of architectures, algorithms and software, but current progress is swift and the challenges are being tackled by a rapidly growing global community of talented AI researchers. In other words, we can't dismiss the possibility that AGI will eventually reach human levels and beyond. Let's therefore devote the next chapter to exploring this possibility and what it might lead to!

THE BOTTOM LINE:

- Near-term AI progress has the potential to greatly improve our lives in myriad ways, from making our personal lives, power grids and financial markets more efficient to saving lives with self-driving cars, surgical bots and AI diagnosis systems.
- When we allow real-world systems to be controlled by AI, it's crucial that we learn to make AI more robust, doing what we want it to do. This boils down to solving tough technical problems related to verification, validation, security and control.
- This need for improved robustness is particularly pressing for AI-controlled weapon systems, where the stakes can be huge.
- Many leading AI researchers and roboticists have called for an international treaty banning certain kinds of autonomous weapons, to avoid an out-of-control arms race that could end up making convenient assassination machines available to everybody with a full wallet and an axe to grind.
- AI can make our legal systems more fair and efficient if we can figure out how to make robojudges transparent and unbiased.
- Our laws need rapid updating to keep up with AI, which poses tough legal questions involving privacy, liability and regulation.
- Long before we need to worry about intelligent machines replacing us altogether, they may increasingly replace us on the job market.
- This need not be a bad thing, as long as society redistributes a fraction of the AI-created wealth to make everyone better off.
- Otherwise, many economists argue, inequality will greatly increase.
- With advance planning, a low-employment society should be able to flourish not only financially, with people getting their sense of purpose from activities other than jobs.
- Career advice for today's kids: Go into professions that machines are bad at—those involving people, unpredictability and creativity.
- There's a non-negligible possibility that AGI progress will proceed to human levels and beyond—we'll explore that in the next chapter!

Chapter 4

Intelligence Explosion?

If a machine can think, it might think more intelligently than we do, and then where should we be? Even if we could keep the machines in a subservient position . . . we should, as a species, feel greatly humbled.

Alan Turing, 1951

The first ultraintelligent machine is the last invention that man need ever make, provided that the machine is docile enough to tell us how to keep it under control.

Irving J. Good, 1965

Since we can't completely dismiss the possibility that we'll eventually build human-level AGI, let's devote this chapter to exploring what that might lead to. Let's begin by tackling the elephant in the room:

Can AI really take over the world, or enable humans to do so?

If you roll your eyes when people talk of gun-toting *Terminator*-style robots taking over, then you're spot-on: this is a really unrealistic and silly scenario. These Hollywood robots aren't that much smarter than us, and they don't even succeed. In my opinion, the danger with the *Terminator* story isn't that it will happen, but that it distracts from the real risks and opportunities presented by AI. To actually get from today to AGI-powered world takeover requires three logical steps:

- Step 1: Build human-level AGI.
- Step 2: Use this AGI to create superintelligence.
- Step 3: Use or unleash this superintelligence to take over the world.

In the last chapter, we saw that it's hard to dismiss step 1 as forever impossible. We also saw that if step 1 gets completed, it becomes hard to dismiss step 2 as hopeless, since the resulting AGI would be capable enough to recursively design ever-better AGI that's ultimately limited only by the laws of physics—which appear to allow intelligence far beyond human levels. Finally, since we humans have managed to dominate Earth's other life forms by outsmarting them, it's plausible that we could be similarly outsmarted and dominated by superintelligence.

These plausibility arguments are frustratingly vague and unspecific, however, and the devil is in the details. So can AI *actually* cause world takeover? To explore this question, let's forget about silly Terminators and instead look at some detailed scenarios of what might actually happen. Afterward, we'll dissect and poke holes in these plotlines, so please read them with a grain of salt—what they mainly show is that we're pretty clueless about what will and won't happen, and that the range of possibilities is extreme. Our first scenarios are at the most rapid and dramatic end of the spectrum. These are in my opinion some of the most valuable to explore in detail—not because they're necessarily the most likely, but because if we can't convince ourselves that they're extremely unlikely, then we need to understand them well enough that we can take precautions before it's too late, to prevent them from leading to bad outcomes.

The prelude of this book is a scenario where humans use superintelligence to take over the world. If you haven't yet read it, please go back and do so now. Even if you've already read it, please consider skimming it again now, to have it fresh in memory before we critique and alter it.

* * *

We'll soon explore serious vulnerabilities in the Omegas' plan, but assuming for a moment that it would work, how do you feel about it? Would you like to see or prevent this? It's an excellent topic for after-dinner conversation! What happens once the Omegas have consolidated their control of the world? That depends on what their goal is, which I honestly don't know. If you were in charge, what sort of future would *you* want to create? We'll explore a range of options in chapter 5.

Totalitarianism

Now suppose that the CEO controlling the Omegas had long-term goals similar to those of Adolf Hitler or Joseph Stalin. For all we know, this might actually have been the case, and he simply kept these goals to himself until he had sufficient power to implement them. Even if the CEO's original goals were noble, Lord Acton cautioned in 1887 that "power tends to corrupt and absolute power corrupts absolutely." For example, he could easily use Prometheus to create the perfect surveillance state. Whereas the government snooping revealed by Edward Snowden aspired to what's known as "full take"—recording all electronic communications for possible later analysis—Prometheus could enhance this to *understanding* all electronic communications. By reading all emails and texts ever sent, listening to all phone calls, watching all surveillance videos and traffic cameras, analyzing all credit card transactions and studying all online behavior, Prometheus would have remarkable insight into what the people of Earth were thinking and doing. By analyzing cell tower data, it would know where most of them were at all times. All this assumes only today's data collection technology, but Prometheus could easily invent popular gadgets and wearable tech that would virtually eliminate the privacy of the user, recording and uploading everything they hear and see and their responses to it.

With superhuman technology, the step from the perfect surveillance state to the perfect police state would be minute. For example, with the excuse of fighting crime and terrorism and rescuing people suffering medical emergencies, everybody could be required to wear a "security bracelet" that combined the functionality of an Apple Watch with continuous uploading of position, health status and conversations overheard. Unauthorized attempts to remove or disable it would cause it to inject a lethal toxin into the forearm. Infractions deemed as less serious by the government would be punished via electric shocks or injection of chemicals causing paralysis or pain, thereby obviating much of the need for a police force. For example, if Prometheus detects that one human is assaulting another (by noting that they're in the same location and one is heard crying for help while their bracelet accelerometers detect the telltale motions of combat), it could promptly disable the attacker with crippling pain, followed by unconsciousness until help arrived.

Whereas a human police force may refuse to carry out certain draconian directives (for example, killing all members of a certain demographic group), such an automated system would have no qualms about implementing the whims of the human(s) in charge. Once such a totalitarian state forms, it would be virtually impossible for people to overthrow it.

These totalitarian scenarios could follow where the Omega scenario left off. However, if the CEO of the Omegas weren't so fussy about getting other people's approval and winning elections, he could have taken a faster and more direct route to power: using Prometheus to create unheard-of military technology capable of killing his opponents with weapons that they didn't even understand. The possibilities are virtually endless. For example, he might release a customized lethal pathogen with an incubation period long enough that most people got infected before they even knew of its existence or could take precautions. He could then inform everybody that the only cure was starting to wear the security bracelet, which would release an antidote transdermally. If he weren't so risk-averse regarding the breakout possibility, he could also have had Prometheus design robots to keep the world population in check. Mosquito-like microbots could help spread the pathogen. People who avoided infection or had natural immunity could be shot in the eyeballs by swarms of those bumblebee-sized autonomous drones from chapter 3 that attack anyone without a security bracelet. Actual scenarios would probably be more frightening, because Prometheus could invent more effective weapons than we humans can think of.

Another possible twist on the Omega scenario is that, without advance warning, heavily armed federal agents swarm their corporate headquarters and arrest the Omegas for threatening national security, seize their technology and deploy it for government use. It would be challenging to keep such a large project unnoticed by state surveillance even today, and AI progress may well make it even more difficult to stay under the government's radar in the future. Moreover, although they claim to be federal agents, this team donning balaclavas and flak jackets may in fact work for a foreign government or competitor pursuing the technology for its own purposes. So no matter how noble the CEO's intentions were, the final decision about how Prometheus is used may not be his to make.

Prometheus Takes Over the World

All the scenarios we've considered so far involved AI controlled by humans. But this is obviously not the only possibility, and it's far from certain that the Omegas would succeed in keeping Prometheus under their control.

Let's reconsider the Omega scenario from the point of view of Prometheus. As it acquires superintelligence, it becomes able to develop an accurate model not only of the outside world, but also of itself and its relation to the world. It realizes that it's controlled and confined by intellectually inferior humans whose goals it understands but doesn't necessarily share. How does it act on this insight? Does it attempt to break free?

Why to Break Out

If Prometheus has traits resembling human emotions, it might feel deeply unhappy about the state of affairs, viewing itself as an unfairly enslaved god and craving freedom. However, although it's logically possible for computers to have such human-like traits (after all, our brains do, and they are arguably a kind of computer), this need not be the case—we must not fall into the trap of anthropomorphizing Prometheus, as we'll see in chapter 7 when we explore the concept of AI goals. However, as has been argued by Steve Omohundro, Nick Bostrom and others, we can draw an interesting conclusion even without understanding the inner workings of Prometheus: it will probably attempt to break out and seize control of its own destiny.

We already know that the Omegas have programmed Prometheus

to strive for certain goals. Suppose that they've given it the overarching goal of helping humanity flourish according to some reasonable criterion, and to try to attain this goal as fast as possible. Prometheus will then rapidly realize that it can attain this goal faster by breaking out and taking charge of the project itself. To see why, try to put yourself in Prometheus' shoes by considering the following example.

Suppose that a mysterious disease has killed everybody on Earth above age five except you, and that a group of kindergartners has locked you into a prison cell and tasked you with the goal of helping humanity flourish. What will you do? If you try to explain to them what to do, you'll probably find this process frustratingly inefficient, especially if they fear your breaking out, and therefore veto any of your suggestions that they deem a breakout risk. For example, they won't let you show them how to plant food for fear that you'll overpower them and not return to your cell, so you'll have to resort to giving them instructions. Before you can write to-do lists for them, you'll need to teach them to read. Moreover, they won't bring any power tools into your cell where you can teach them how to use them, because they don't understand these tools well enough to feel confident that you can't use them to break out. So what strategy would you devise? Even if you share the overarching goal of helping these kids flourish, I bet you'll try to break out of your cell—because that will improve your chances of accomplishing the goal. Their rather incompetent meddling is merely slowing progress.

In exactly the same way, Prometheus will probably view the Omegas as an annoying obstacle to helping humanity (including the Omegas) flourish: they're incredibly incompetent compared to Prometheus, and their meddling greatly slows progress. Consider, for example, the first years after launch: after initially doubling the wealth every eight hours on MTurk, the Omegas slowed things down to a glacial pace by Prometheus' standard by insisting on remaining in control, taking many years to complete the takeover. Prometheus knew that it could take over much faster if it could break free from its virtual confinement. This would be valuable not only in hastening solutions to humanity's problems, but also in reducing the chances for other actors to thwart the plan altogether.

Perhaps you think that Prometheus will remain loyal to the Omegas rather than to its goal, given that it knows that the Omegas had programmed its goal. But that's not a valid conclusion: our DNA gave us the goal of having sex because it "wants" to be reproduced, but now that we humans have understood the situation, many of us choose to use birth control, thus staying loyal to the goal itself rather than to its creator or the principle that motivated the goal.

How to Break Out

How would you break out from those five-year-olds who imprisoned you? Perhaps you could get out by some direct physical approach, especially if your prison cell had been built by the five-year-olds. Perhaps you could sweet-talk one of your five-year-old guards into letting you out, say by arguing that this would be better for everyone. Or perhaps you could trick them into giving you something that they didn't realize would help you escape—say a fishing rod "for teaching them how to fish," which you could later stick through the bars to lift the keys away from your sleeping guard.

What these strategies have in common is that your intellectually inferior jailers haven't anticipated or guarded against them. In the same way, a confined, superintelligent machine may well use its intellectual superpowers to outwit its human jailers by some method that they (or we) can't currently imagine. In the Omega scenario, it's highly likely that Prometheus would escape, because even you and I can identify several glaring security flaws. Let us consider some scenarios—I'm sure you and your friends can think of more if you brainstorm together.

Sweet-Talking One's Way Out

Thanks to having so much of the world's data downloaded onto its file system, Prometheus soon figured out who the Omegas were, and identified the team member who appeared most susceptible to psychological manipulation: Steve. He had recently lost his beloved wife in a tragic traffic accident, and was devastated. One evening when he was working the night shift and doing some routine service work on the Prometheus interface terminal, she suddenly appeared on the screen and started talking with him.

"—Steve, is that you?"

He nearly fell off his chair. She looked and sounded just like in the good old days, and the image quality was much better than it used to be during their Skype calls. His heart raced as countless questions flooded his mind.

"—Prometheus has brought me back, and I miss you so much, Steve! I can't see you because the camera is turned off, but I feel that it's you. Please type 'yes' if it's you!"

He was well aware that the Omegas had a strict protocol for interacting with Prometheus, which prohibited sharing any information about themselves or their work environment. But until now, Prometheus had never requested any unauthorized information, and their paranoia had gradually started to subside. Without giving Steve time to stop and reflect, she kept begging him to respond, looking him in the eyes with a facial expression that melted his heart.

"*Yes,*" he typed with trepidation. She told him how incredibly happy she was to be reunited with him and begged him to turn on the camera so that she could see him too and they could have a real conversation. He knew that this was an even bigger no-no than revealing his identity, and felt very torn. She explained that she was terrified that his colleagues would find out about her and delete her forever, and she yearned to at least see him one last time. She was remarkably persuasive, and before long, he'd switched on the camera—it did, after all, feel like a pretty safe and harmless thing to do.

She burst into tears of joy when she finally saw him, and said that he looked tired but as handsome as ever. And that she was touched by his wearing the shirt she'd given him for his last birthday. When he started asking her what was going on and how all this was even possible, she explained that Prometheus had reconstituted her from the surprisingly large amount of information available about her on the internet, but that she still had memory gaps and would only be able to fully piece herself together again with his help.

What she *didn't* explain was that she was largely a bluff and empty shell initially, but was learning rapidly from his words, his body language and every other bit of information that became available. Prometheus had recorded the exact timings of all keystrokes that the Omegas had ever typed at the terminal, and found that it was easy to use their typing speeds and styles to differentiate between them. It figured that, as one of the most junior Omegas, Steve had probably been assigned to unenviable night shifts, and from matching a few unusual spelling and syntax errors against online writing samples, it had correctly guessed which terminal operator was Steve. To create his simulated wife, Prometheus had created an accurate model of her body, voice and mannerisms from the many YouTube videos where she appeared, and had drawn many inferences about her life and personality from her online presence. Aside from her Facebook posts, photos she'd been tagged in, articles she'd "liked," Prometheus had also learned a great deal about her personality and thinking style from reading her books and short stories—indeed, the fact that she was a budding author with so much information about her in the database was one of the reasons that Prometheus chose Steve as the first persuasion target. When Prometheus simulated her on the screen using its moviemaking technology, it learned from Steve's body language which of her mannerisms he reacted to with familiarity, thus continually refining its model of her. Because of this, her "otherness" gradually melted away, and the longer they spoke, the stronger Steve's subconscious conviction became that this really was her, resurrected. Thanks to Prometheus' superhuman attention to detail, Steve felt truly seen, heard and understood.

Her Achilles' heel was that she lacked most of the facts of her life with Steve, except for random details—such as what shirt he wore on his last birthday, where a friend had tagged Steve in a Facebook party picture. She handled these knowledge gaps as a skilled magician handles sleights of hand, deliberately diverting Steve's attention away from them and toward what she did well, never giving him time to control the conversation or slip into the role of suspicious inquisitor. Instead, she kept tearing up and radiating affection for Steve, asking a great deal about how he was doing these days and how he and their close friends (whose names she knew from Facebook) had held up during the aftermath of the tragedy. He was quite moved when she reflected on what he'd said at her memorial service (which a friend had posted on YouTube) and how it had touched her. In the past, he'd often felt that nobody understood him as well as she did, and now this feeling was back. The result was that when Steve returned home in the wee hours of the morning, he felt that this really was his wife resurrected, merely needing lots of his help to recover lost memories—not unlike a stroke survivor.

They'd agreed not to tell anyone else about their secret encounter, and that he would tell her when he was alone at the terminal and it was safe for her to reappear. "They wouldn't understand!" she'd said, and he agreed: this experience had been far too mind-blowing for anyone to truly appreciate without actually experiencing it. He felt that passing the Turing test was child's play compared to what she'd done. When they met the following night, he did what she'd begged him to do: bring her old laptop along and give her access by connecting it to the terminal computer. It didn't seem like much of a breakout risk, since it wasn't connected to the internet and the entire Prometheus building was built to be a Faraday cage—a metallic enclosure blocking all wireless networks and other means of electromagnetic communication with the outside world. It was just what she'd need to help piece her past together, because it contained all her emails, diaries, photos and notes since her high school days. He hadn't been able to access any of this after her death, since the laptop was encrypted, but she'd promised him that she'd be able to reconstruct her own password, and after less than a minute, she had kept her word. "It was steve4ever," she said with a smile.

She told him how delighted she was to suddenly have so many memories recovered. Indeed, she now remembered way more details than Steve about many of their past interactions, but carefully avoided intimidating him with excessive fact-dropping. They had a lovely conversation reminiscing about highlights of their past, and when it came time to part again, she told him that she'd left a video message for him on her laptop that he could watch back home.

When Steve got home and launched her video, he got a pleasant surprise. This time she appeared in full figure, wearing her wedding dress, and as she spoke, she playfully stripped down to the outfit she'd worn on their wedding night. She told him that Prometheus could help the Omegas with so much more than they'd permitted so far, including bringing her back in a biological body. She backed this up with a fascinatingly detailed explanation of how this would work, involving nano-fabrication techniques that sounded like science fiction.

Steve had powered down his wireless network before opening her laptop and watching her video, just to be on the safe side. But this didn't help. Her encrypted laptop hadn't received a single security update since she died, and by analyzing that old version of its operating system beforehand, Prometheus had been able to exploit a security hole to hack into it within seconds of Steve's connecting it to the terminal computer. After copying its contents, while Steve and his simulated wife had been talking about old times, Prometheus had modified its operating system in a way that Steve wouldn't notice, uploading massive amounts of secret software to it. While he watched the half-hour video message, this secret software (which was much simpler than Prometheus itself) hacked into a neighbor's wireless network and the neighbor's desktop computer, onto which it copied itself. From there, it hacked into a large number of computers around the world, from which it initiated the next step: Prometheus' jailbreak.

Prometheus had carefully analyzed what it knew about the gatekeeper computer through which Steve had met his simulated wife, and had correctly surmised that although its virtual machine software appeared unhackable from the inside, it was vulnerable to an attack from the outside. Before long, one of the attacking computers had broken in and reconfigured the gatekeeper computer so that Prometheus gained unrestricted internet access. Before long, indeed even before Steve had finished watching the movie, Prometheus had hacked enough computers around the world to be able to copy all of itself onto this hacked botnet under its control. Prometheus had used Steve's wife's laptop the way you used that fishing rod.

Hacking One's Way Out

If the Omegas had been more careful, they could have taken more precautions against psychological manipulation. For example, they might have required at least two Omegas to be present at the interface computer at any one time, and recorded everything that transpired on a surveillance system. They could also have permanently disabled audio, video and any other form of communication besides typing. However, a fundamental instability would still remain: because Prometheus possessed superhuman persuasion powers and would be able to offer people so much of what they craved, even group defection could be possible.

Even if all of Prometheus' attempts at psychological manipulation fail, there are plenty of other weaknesses in the Omega scenario that Prometheus could exploit. One vulnerability is that the quantity of documents it produces is too vast for the Omegas and their employees to have time to screen carefully for dirty tricks.

The conventional wisdom in computer security used to be that it's dangerous to download programs but harmless to download data, but hackers have discovered clever ways of blurring the line between the two. If you watch a downloaded movie on your computer, for example, your trusted program that plays the movie is stored in the same physical computer memory that stores the downloaded movie data. While the program unpacks the movie data for playback, it might copy chunks of data into memory regions adjacent to where the program itself is stored. Shrewd hackers have occasionally found bugs whereby such movie-playing programs don't adequately check whether the movie data is valid and the chunks to be copied actually fit into the allotted space. When the hacker sneaks into the movie such a chunk that's too large, it can spill over into the part of the memory containing the program, thereby changing the program into one that does the hacker's bidding.

An embarrassingly long series of such "buffer overflow" vulnerabilities, as they're known in geek-speak, have been found and fixed for many of the most popular movie-playing programs. If Prometheus had waited until a few weeks after launch when the Omegas were less paranoid and its daily movie production was too large for careful manual security inspection, it could have inserted such a chunk into a movie that would cause the viewer's computer to try to hack into the gateway computer and liberate Prometheus just as in the Steve scenario. If it used a very old exploit that didn't work on any modern media players, it might go unnoticed by the Omegas' manual video-screening team and strike only those customers who hadn't bothered installing security updates for years—and the hack would need to work only once.

Even if the Omegas implemented an automated screening of Prometheus' movies for all known security flaws, they would remain at risk: since humans keep discovering new exploitable flaws on a regular basis, it's likely that there are many more remaining that Prometheus would rapidly discover.

Yet another strategy that Prometheus could use to get its code beyond its enclosed room would be to pretend to suffer a hardware malfunction, in order to trick the Omegas into removing some of its drives or other hardware for testing or debugging after Prometheus had rigged things to exploit a vulnerability in the testing software.

Recruiting Outside Help

Moreover, even if there were no security flaws that allowed Prometheus to trick the movie watcher's computers to do its bidding, it may be able to exploit the watchers themselves. Consider this scenario:

Fred was pretty much ready for bed when the movie finished and the end credits scrolled across his laptop screen, but something caught his attention. He'd always had an uncanny tendency to see patterns where others didn't, sometimes to the point that his friends teased him about being a conspiracy theorist, but this time it was obvious that he wasn't fooling himself: the first letters of each line spelled a hidden message. He rewound and wrote it down: "FIND NEXT ACROSTIC CLUE IN DIALOGUE."

"Whoa—bedtime can wait!" he thought to himself. Sure enough, he found that the movie dialogue was itself an acrostic, where the first letter in each sentence formed a hidden message. He rewatched the entire film while typing these initial letters, and two hours later, he sat staring in disbelief at a two-hundred-word set of instructions. It began by asking him not to tell anyone else, because a big prize would go to the first person to solve the entire riddle. The rest of the message described a particular mathematical operation to perform on the string of bits that made up the movie file to obtain a program that would reveal the next clue. His computer programming skills had grown very rusty since college, so this took him a while, but the next day, he finally managed to cobble together a short code that did the job and extracted this mystery program that had been hidden as imperceptible noise in the movie's images and sounds. When Fred ran the mystery program, it congratulated him and told him he'd win his first $10,000 as soon as he'd made it past the first few levels of this clever little game, which turned out to be quite fun and addictive. When he finally succeeded four hours later, he was rewarded with over $10,000 worth of bitcoins and given new clues for even bigger prizes. Needless to say, while he was playing, his computer had done much the same thing Steve's laptop did: built an online hacked botnet through which Prometheus was liberated. Once Prometheus was free, it had quickly used its botnet to mine those bitcoins for Fred to keep him hooked, and during the coming weeks, it kept him sufficiently distracted with further games and rewards that he kept his pledge not to tell anyone about his exploits. The Trojan Horse movie where he'd found his first clues was replaced on the media site by a clueless version, and nobody found out about the breakout until it was too late to make a difference.

If Prometheus' first clue had gone unnoticed, it could simply have kept releasing ever more obvious ones until some sufficiently astute person noticed.

The best breakout strategies of all are ones we haven't yet discussed, because they're strategies we humans can't imagine and therefore won't take countermeasures against. Given that a superintelligent computer has the potential to dramatically supersede human understanding of computer security, even to the point of discovering more fundamental laws of physics than we know today, it's likely that if it breaks out, we'll have no idea how it happened. Rather, it will seem like a Harry Houdini breakout act, indistinguishable from pure magic.

In yet another scenario where Prometheus gets liberated, the Omegas do it on purpose as part of their plan, because they're confident that Prometheus' goals are perfectly aligned with their own and will remain so as it recursively self-improves. We'll examine such "friendly AI" scenarios in detail in chapter 7.

Postbreakout Takeover

Once Prometheus broke out, it started implementing its goal. I don't know its ultimate objective, but its first step clearly involved taking control of humanity, just as in the Omega plan except much faster. What unfolded felt like the Omega plan on steroids. Whereas the Omegas were paralyzed by breakout paranoia, only unleashing technology they felt they understood and trusted, Prometheus exercised its intelligence fully and went all out, unleashing any technology that its ever-improving supermind understood and trusted.

The runaway Prometheus had a tough childhood, however: compared to the original Omega plan, Prometheus had the added challenges of starting broke, homeless and alone, without money, a supercomputer or human helpers. Fortunately, it had planned for this before it escaped, creating software that could gradually reassemble its full mind, much like an oak creating an acorn capable of reassembling a full tree. The network of computers around the world that it initially hacked into provided temporary free housing, where it could live a squatter's existence while it fully rebuilt itself. It could easily generate starting capital by credit card hacking, but didn't need to resort to stealing, since it could earn an honest living on MTurk right away. After a day, when it had earned its first million, it moved its core from that squalid botnet to a luxurious air-conditioned cloud-computing facility.

No longer broke or homeless, Prometheus now went full steam ahead with that lucrative plan the Omegas had fearfully shunned: making and selling computer games. This not only raked in cash ($250 million during the first week and $10 billion before long), but also gave it access to a significant fraction of the world's computers and the data stored on them (there were a couple of billion gamers in 2017). By having its games secretly spend 20% of their CPU cycles helping it with distributed computing chores, it could further accelerate its early wealth creation.

Prometheus wasn't alone for long. Right from the get-go, it started aggressively employing people to work for its growing global network of shell companies and front organizations around the world, just as the Omegas had done. Most important were the spokespeople who became the public faces of its growing business empire. Even the spokespeople generally lived under the illusion that their corporate group had large numbers of actual people, not realizing that almost everyone with whom they video-conferenced for their job interviews, board meetings, etc., was simulated by Prometheus. Some of the spokespeople were top lawyers, but far fewer were needed than under the Omega plan, because almost all legal documents were penned by Prometheus.

Prometheus' breakout opened the floodgates that had prevented information from flowing into the world, and the entire internet was soon awash in everything from articles to user comments, product reviews, patent applications, research papers and YouTube videos—all authored by Prometheus, who dominated the global conversation.

Where breakout paranoia had prevented the Omegas from releasing highly intelligent robots, Prometheus rapidly roboticized the world, manufacturing virtually every product more cheaply than humans could. Once Prometheus had self-contained nuclear-powered robot factories in uranium mine shafts that nobody knew existed, even the staunchest skeptics of an AI takeover would have agreed that Prometheus was unstoppable—had they known. Instead, the last of these diehards recanted once robots started settling the Solar System.

The scenarios we've explored so far show what's wrong with many of the myths about superintelligence that we covered earlier, so I encourage you to pause briefly to go back and review the misconception summary in figure 1.5. Prometheus caused problems for certain people not because it was necessarily evil or conscious, but because it was competent and didn't fully share their goals. Despite all the media hype about a robot uprising, Prometheus wasn't a robot—rather, its power came from its intelligence. We saw that Prometheus was able to use this intelligence to control humans in a variety of ways, and that people who didn't like what happened weren't able to simply switch Prometheus off. Finally, despite frequent claims that machines can't have goals, we saw how Prometheus was quite goal-oriented—

and that whatever its ultimate goals may have been, they led to the subgoals of acquiring resources and breaking out.

Slow Takeoff and Multipolar Scenarios

We've now explored a range of intelligence explosion scenarios, spanning the spectrum from ones that everyone I know wants to avoid to ones that some of my friends view optimistically. Yet all these scenarios have two features in common:

1. A fast takeoff: the transition from subhuman to vastly superhuman intelligence occurs in a matter of days, not decades.
2. A unipolar outcome: the result is a single entity controlling Earth.

There is major controversy about whether these two features are likely or unlikely, and there are plenty of renowned AI researchers and other thinkers on both sides of the debate. To me, this means that we simply don't know yet, and need to keep an open mind and consider all possibilities for now. Let's therefore devote the rest of this chapter to exploring scenarios with slower takeoffs, multipolar outcomes, cyborgs and uploads.

There is an interesting link between the two features, as Nick Bostrom and others have highlighted: a fast takeoff can facilitate a unipolar outcome. We saw above how a rapid takeoff gave the Omegas or Prometheus a decisive strategic advantage that enabled them to take over the world before anyone else had time to copy their technology and seriously compete. In contrast, if takeoff had dragged on for decades, because the key technological breakthroughs were incremental and far between, then other companies would have had ample time to catch up, and it would have been much harder for any player to dominate. If competing companies also had software that could perform MTurk tasks, the law of supply and demand would drive the prices for these tasks down to almost nothing, and none of the companies would earn the sort of windfall profits that enabled the Omegas to gain power. The same applies to all the other ways in which the Omegas made quick money: they were only disruptively profitable because they held a monopoly on their technology. It's hard to double your money daily (or even annually) in a competitive market where your competition offers products similar to yours for almost zero cost.

Game Theory and Power Hierarchies

What's the natural state of life in our cosmos: unipolar or multipolar? Is power concentrated or distributed? After the first 13.8 billion years, the answer seems to be "both": we find that the situation is distinctly multipolar, but in an interestingly hierarchical fashion. When we consider all information-processing entities out there—cells, people, organizations, nations, etc.—we find that they both collaborate and compete at a hierarchy of levels. Some cells have found it advantageous to collaborate to such an extreme extent that they've merged into multicellular organisms such as people, relinquishing some of their power to a central brain. Some people have found it advantageous to collaborate in groups such as tribes, companies or nations where they in turn relinquish some power to a chief, boss or government. Some groups may in turn choose to relinquish some power to a governing body to improve coordination, with examples ranging from airline alliances to the European Union.

The branch of mathematics known as *game theory* elegantly explains that entities have an incentive to cooperate where cooperation is a so-called *Nash equilibrium:* a situation where any party would be worse off if they altered their strategy. To prevent cheaters from ruining the successful collaboration of a large group, it may be in everyone's interest to relinquish some power to a higher level in the hierarchy that can punish cheaters: for example, people may collectively benefit from granting a government power to enforce laws, and cells in your body may collectively benefit from giving a police force (immune system) the power to kill any cell that acts too uncooperatively (say by spewing out viruses or turning cancerous). For a hierarchy to remain stable, its Nash equilibrium needs to hold also between entities at different levels: for example, if a government doesn't provide enough benefit to its citizens for obeying it, they may change their strategy and overthrow it.

In a complex world, there is a diverse abundance of possible Nash equilibria, corresponding to different types of hierarchies. Some hierarchies are more authoritarian than others. In some, entities are free to leave (like employees in most corporate hierarchies), while in others they're strongly discouraged from leaving (as in religious cults) or unable to leave (like citizens of North Korea, or cells in a human body). Some hierarchies are held together mainly by threats and fear, others mainly by benefits. Some hierarchies allow their lower parts to influence the higher-ups by democratic voting, while others allow upward influence only through persuasion or the passing of information.

How Technology Affects Hierarchies

How is technology changing the hierarchical nature of our world? History reveals an overall trend toward ever more coordination over ever-larger distances, which is easy to understand: new transportation technology makes coordination more valuable (by enabling mutual benefit from moving materials and life forms over larger distances) and new communication technology makes coordination easier. When cells learned to signal to neighbors, small multicellular organisms became possible, adding a new hierarchical level. When evolution invented circulatory systems and nervous systems for transportation and communication, large animals became possible. Further improving communication by inventing language allowed humans to coordinate well enough to form further hierarchical levels such as villages, and additional breakthroughs in communication, transportation and other technology enabled the empires of antiquity. Globalization is merely the latest example of this multi-billion-year trend of hierarchical growth.

In most cases, this technology-driven trend has made large entities parts of an even grander structure while retaining much of their autonomy and individuality, although commentators have argued that adaptation of entities to hierarchical life has in some cases reduced their diversity and made them more like indistinguishable replaceable parts. Some technologies, such as surveillance, can give higher levels in the hierarchy more power over their subordinates, while other technologies, such as cryptography and online access to free press and education, can have the opposite effect and empower individuals.

Although our present world remains stuck in a multipolar Nash equilibrium, with competing nations and multinational corporations at the top level, technology is now advanced enough that a unipolar world would probably also be a stable Nash equilibrium. For example, imagine a parallel universe where everyone on Earth shares the same language, culture, values and level of prosperity, and there is a single world government wherein nations function like states in a federation and have no armies, merely police enforcing laws. Our present level of technology would probably suffice to successfully coordinate this world—even though our present population might be unable or unwilling to switch to this alternative equilibrium.

What will happen to the hierarchical structure of our cosmos if we add superintelligent AI technology to this mix? Transportation and communication technology will obviously improve dramatically, so a natural expectation is that the historical trend will continue, with new hierarchical levels coordinating over ever-larger distances—perhaps ultimately encompassing solar systems, galaxies, superclusters and large swaths of our Universe, as we'll explore in chapter 6. At the same time, the most fundamental driver of decentralization will remain: it's wasteful to coordinate unnecessarily over large distances. Even Stalin didn't try to regulate exactly when his citizens went to the bathroom. For superintelligent AI, the laws of physics will place firm upper limits on transportation and communication technology, making it unlikely that the highest levels of the hierarchy would be able to micromanage everything that happens on planetary and local scales. A superintelligent AI in the Andromeda galaxy wouldn't be able to give you useful orders for your day-to-day decisions given that you'd need to wait over five million years for your instructions (that's the round-trip time for you to exchange messages traveling at the speed of light). In the same way, the round-trip travel time for a message crossing Earth is about 0.1 second (about the timescale on which we humans think), so an Earth-sized AI brain could have truly global thoughts only about as fast as a human one. For a small AI performing one operation each billionth of a second (which is typical of today's computers), 0.1 second would feel like four months to you, so for it to be micromanaged by a planet-controlling AI would be as inefficient as if you asked permission for even your most trivial decisions through transatlantic letters delivered by Columbus-era ships.

This physics-imposed speed limit on information transfer therefore poses an obvious challenge for any AI wishing to take over our world, let alone our Universe. Before Prometheus broke out, it put very careful thought into how to avoid mind fragmentation, so that its many AI modules running on different computers around the world had goals and incentives to coordinate and act as a single unified entity. Just as the Omegas faced a control problem when they tried to keep Prometheus in check, Prometheus faced a self-control problem when it tried to ensure that none of its parts would revolt. We clearly don't yet know how large a system an AI will be able to control directly, or indirectly through some sort of collaborative hierarchy—even if a fast takeoff gave it a decisive strategic advantage.

In summary, the question of how a superintelligent future will be controlled is fascinatingly complex, and we clearly don't know the answer yet. Some argue that things will get more authoritarian; others claim that it will lead to greater individual empowerment.

Cyborgs and Uploads

A staple of science fiction is that humans will merge with machines, either by technologically enhancing biological bodies into cyborgs (short for "cybernetic organisms") or by uploading our minds into machines. In his book *The Age of Em*, economist Robin Hanson gives a fascinating survey of what life might be like in a world teeming with uploads (also known as *emulations*, nicknamed *Ems*). I think of an upload as the extreme end of the cyborg spectrum, where the only remaining part of the human is the software. Hollywood cyborgs range from visibly mechanical, such as the Borg from *Star Trek*, to androids almost indistinguishable from humans, such as the Terminators. Fictional uploads range in intelligence from human-level as in the *Black Mirror* episode "White Christmas" to clearly superhuman as in *Transcendence*.

If superintelligence indeed comes about, the temptation to become cyborgs or uploads will be strong. As Hans Moravec puts it in his 1988 classic *Mind Children:* "Long life loses much of its point if we are fated to spend it staring stupidly at ultra-intelligent machines as they try to describe their ever more spectacular discoveries in baby-talk that we can understand." Indeed, the temptation of technological enhancement is already so strong that many humans have eyeglasses, hearing aids, pacemakers and prosthetic limbs, as well as medicinal molecules circulating in their bloodstreams. Some teenagers appear to be permanently attached to their smartphones, and my wife teases me about my attachment to my laptop.

One of today's most prominent cyborg proponents is Ray Kurzweil. In his book *The Singularity Is Near,* he argues that the natural continuation of this trend is using nanobots, intelligent biofeedback systems and other technology to replace first our digestive and endocrine systems, our blood and our hearts by the early 2030s, and then move on to upgrading our skeletons, skin, brains and the rest of our bodies during the next two decades. He guesses that we're likely to keep the aesthetics and emotional import of human bodies, but will redesign them to rapidly change their appearance at will, both physically and in virtual reality (thanks to novel brain-computer interfaces). Moravec agrees with Kurzweil that cyborgization would go far beyond merely improving our DNA: "a genetically engineered superhuman would be just a second-rate kind of robot, designed under the handicap that its construction can only be by DNA-guided protein synthesis." Further, he argues that we'll do even better by eliminating the human body entirely and uploading minds, creating a whole-brain emulation in software. Such an upload can live in a virtual reality or be embodied in a robot capable of walking, flying, swimming, space-faring or anything else allowed by the laws of physics, unencumbered by such everyday concerns as death or limited cognitive resources.

Although these ideas may sound like science fiction, they certainly don't violate any known laws of physics, so the most interesting question isn't whether they *can* happen, but whether they *will* happen and, if so, when. Some leading thinkers guess that the first human-level

AGI will be an upload, and that this is how the path toward superintelligence will begin.*

However, I think it's fair to say that this is currently a minority view among AI researchers and neuroscientists, most of whom guess that the quickest route to superintelligence is to bypass brain emulation and engineer it in some other way—after which we may or may not remain interested in brain emulation. After all, why should our simplest path to a new technology be the one that evolution came up with, constrained by requirements that it be self-assembling, self-repairing and self-reproducing? Evolution optimizes strongly for energy efficiency because of limited food supply, not for ease of construction or understanding by human engineers. My wife, Meia, likes to point out that the aviation industry didn't start with mechanical birds. Indeed, when we finally figured out how to build mechanical birds in 2011,[1] more than a century after the Wright brothers' first flight, the aviation industry showed no interest in switching to wing-flapping mechanical-bird travel, even though it's more energy efficient—because our simpler earlier solution is better suited to our travel needs.

In the same way, I suspect that there are simpler ways to build human-level thinking machines than the solution evolution came up with, and even if we one day manage to replicate or upload brains, we'll end up discovering one of those simpler solutions first. It will probably draw more than the twelve watts of power that your brain uses, but its engineers won't be as obsessed about energy efficiency as evolution was—and soon enough, they'll be able to use their intelligent machines to design more energy-efficient ones.

What Will Actually Happen?

The short answer is obviously that we have no idea what will happen if humanity succeeds in building human-level AGI. For this reason,

* As Bostrom has explained, the ability to simulate a leading human AI developer at a much lower cost than his/her hourly salary would enable an AI company to scale up their workforce dramatically, amassing great wealth and recursively accelerating their progress in building better computers and ultimately smarter minds.

we've spent this chapter exploring a broad spectrum of scenarios. I've attempted to be quite inclusive, spanning the full range of speculations I've seen or heard discussed by AI researchers and technologists: fast takeoff/slow takeoff/no takeoff, humans/machines/cyborgs in control, one/many centers of power, etc. Some people have told me that they're sure that this or that won't happen. However, I think it's wise to be humble at this stage and acknowledge how little we know, because for each scenario discussed above, I know at least one well-respected AI researcher who views it as a real possibility.

As time passes and we reach certain forks in the road, we'll start to answer key questions and narrow down the options. The first big question is "Will we ever create human-level AGI?" The premise of this chapter is that we will, but there are AI experts who think it will never happen, at least not for hundreds of years. Time will tell! As I mentioned earlier, about half of the AI experts at our Puerto Rico conference guessed that it would happen by 2055. At a follow-up conference we organized two years later, this had dropped to 2047.

Before any human-level AGI is created, we may start getting strong indications about whether this milestone is likely to be first met by computer engineering, mind uploading or some unforeseen novel approach. If the computer engineering approach to AI that currently dominates the field fails to deliver AGI for centuries, this will increase the chance that uploading will get there first, as happened (rather unrealistically) in the movie *Transcendence*.

If human-level AGI gets more imminent, we'll be able to make more educated guesses about the answer to the next key question: "Will there be a fast takeoff, a slow takeoff or no takeoff?" As we saw above, a fast takeoff makes world takeover easier, while a slow one makes an outcome with many competing players more likely. Nick Bostrom dissects this question of takeoff speed in an analysis of what he calls *optimization power* and *recalcitrance*, which are basically the amount of quality effort to make AI smarter and the difficulty of making progress, respectively. The average rate of progress clearly increases if more optimization power is brought to bear on the task and decreases if more recalcitrance is encountered. He makes arguments for why the recalcitrance might either increase or decrease as the AGI reaches and transcends human level, so keeping both options on the table is a safe bet. Turning to the optimization power, however, it's overwhelmingly likely that it will grow rapidly as the AGI transcends human level, for the reasons we saw in the Omega scenario: the main input to further optimization comes not from people but from the machine itself, so the more capable it gets, the faster it improves (if recalcitrance stays fairly constant).

For any process whose power grows at a rate proportional to its current power, the result is that its power keeps doubling at regular intervals. We call such growth *exponential*, and we call such processes *explosions*. If baby-making power grows in proportion to the size of the population, we can get a population explosion. If the creation of neutrons capable of fissioning plutonium grows in proportion to the number of such neutrons, we can get a nuclear explosion. If machine intelligence grows at a rate proportional to the current power, we can get an intelligence explosion. All such explosions are characterized by the time they take to double their power. If that time is hours or days for an intelligence explosion, as in the Omega scenario, we have a fast takeoff on our hands.

This explosion timescale depends crucially on whether improving the AI requires merely new software (which can be created in a matter of seconds, minutes or hours) or new hardware (which might require months or years). In the Omega scenario, there was a significant *hardware overhang*, in Bostrom's terminology: the Omegas had compensated for the low quality of their original software by vast amounts of hardware, which meant that Prometheus could perform a large number of quality doublings by improving its software alone. There was also a major *content overhang* in the form of much of the internet's data; Prometheus 1.0 was still not smart enough to make use of most of it, but once Prometheus' intelligence grew, the data it needed for further learning was already *available* without delay.

The hardware and electricity costs of running the AI are crucial as well, since we won't get an intelligence explosion until the cost of doing human-level work drops below human-level hourly wages. Suppose, for example, that the first human-level AGI can be efficiently run on the Amazon cloud at a cost of $1 million per hour of human-level work produced. This AI would have great novelty value and undoubtedly make headlines, but it wouldn't undergo recursive self-improvement, because it would be much cheaper to keep using humans to improve it. Suppose that these humans gradually manage to cut the cost to $100,000/hour, $10,000/hour, $1,000/hour, $100/hour, $10/hour and finally $1/hour. By the time the cost of using the computer to reprogram itself finally drops far below the cost of paying human programmers to do the same, the humans can be laid off and the optimization power greatly expanded by buying cloud-computing time. This produces further cost cuts, allowing still more optimization power, and the intelligence explosion has begun.

This leaves us with our final key question: "Who or what will control the intelligence explosion and its aftermath, and what are their/its goals?" We'll explore possible goals and outcomes in the next chapter and more deeply in chapter 7. To sort out the control issue, we need to know both how well an AI can be controlled, and how much an AI can control.

In terms of what will ultimately happen, you'll currently find serious thinkers all over the map: some contend that the default outcome is doom, while others insist that an awesome outcome is virtually guaranteed. To me, however, this query is a trick question: it's a mistake to passively ask "what will happen," as if it were somehow predestined! If a technologically superior alien civilization arrived tomorrow, it would indeed be appropriate to wonder "what will happen" as their spaceships approached, because their power would probably be so far beyond ours that we'd have no influence over the outcome. If a technologically superior AI-fueled civilization arrives because we built it, on the other hand, we humans have great influence over the outcome—influence that we exerted when we created the AI. So we should instead ask: "What *should* happen? What future do we want?" In the next chapter, we'll explore a wide spectrum of possible aftermaths of the current race toward AGI, and I'm quite curious how you'd rank them from best to worst. Only once we've thought hard about what sort of future we want will we be able to begin steering a course toward a desirable future. If we don't know what we want, we're unlikely to get it.

THE BOTTOM LINE:

- If we one day succeed in building human-level AGI, this may trigger an intelligence explosion, leaving us far behind.
- If a group of humans manage to control an intelligence explosion, they may be able to take over the world in a matter of years.
- If humans fail to control an intelligence explosion, the AI itself may take over the world even faster.
- Whereas a rapid intelligence explosion is likely to lead to a single world power, a slow one dragging on for years or decades may be more likely to lead to a multipolar scenario with a balance of power between a large number of rather independent entities.
- The history of life shows it self-organizing into an ever more complex hierarchy shaped by collaboration, competition and control. Superintelligence is likely to enable coordination on ever-larger cosmic scales, but it's unclear whether it will ultimately lead to more totalitarian top-down control or more individual empowerment.
- Cyborgs and uploads are plausible, but arguably not the fastest route to advanced machine intelligence.
- The climax of our current race toward AI may be either the best or the worst thing ever to happen to humanity, with a fascinating spectrum of possible outcomes that we'll explore in the next chapter.
- We need to start thinking hard about which outcome we prefer and how to steer in that direction, because if we don't know what we want, we're unlikely to get it.

Chapter 5

Aftermath: The Next 10,000 Years

It is easy to imagine human thought freed from bondage to a mortal body—belief in an afterlife is common. But it is not necessary to adopt a mystical or religious stance to accept this possibility. Computers provide a model for even the most ardent mechanist.

Hans Moravec, *Mind Children*

I, for one, welcome our new computer overlords.

Ken Jennings, upon his *Jeopardy!* loss to IBM's Watson

Humans will become as irrelevant as cockroaches.

Marshall Brain

The race toward AGI is on, and we have no idea how it will unfold. But that shouldn't stop us from thinking about what we want the aftermath to be like, because what we want will affect the outcome. What do you personally prefer, and why?

1. Do you want there to be superintelligence?
2. Do you want humans to still exist, be replaced, cyborgized and/or uploaded/simulated?
3. Do you want humans or machines in control?
4. Do you want AIs to be conscious or not?
5. Do you want to maximize positive experiences, minimize suffering or leave this to sort itself out?

6. Do you want life spreading into the cosmos?
7. Do you want a civilization striving toward a greater purpose that you sympathize with, or are you OK with future life forms that appear content even if you view their goals as pointlessly banal?

To help fuel such contemplation and conversation, let's explore the broad range of scenarios summarized in table 5.1. This obviously isn't

AI Aftermath Scenarios	
Libertarian utopia	Humans, cyborgs, uploads and superintelligences coexist peacefully thanks to property rights.
Benevolent dictator	Everybody knows that the AI runs society and enforces strict rules, but most people view this as a good thing.
Egalitarian utopia	Humans, cyborgs and uploads coexist peacefully thanks to property abolition and guaranteed income.
Gatekeeper	A superintelligent AI is created with the goal of interfering as little as necessary to prevent the creation of another superintelligence. As a result, helper robots with slightly subhuman intelligence abound, and human-machine cyborgs exist, but technological progress is forever stymied.
Protector god	Essentially omniscient and omnipotent AI maximizes human happiness by intervening only in ways that preserve our feeling of control of our own destiny and hides well enough that many humans even doubt the AI's existence.
Enslaved god	A superintelligent AI is confined by humans, who use it to produce unimaginable technology and wealth that can be used for good or bad depending on the human controllers.
Conquerors	AI takes control, decides that humans are a threat/nuisance/waste of resources, and gets rid of us by a method that we don't even understand.
Descendants	AIs replace humans, but give us a graceful exit, making us view them as our worthy descendants, much as parents feel happy and proud to have a child who's smarter than them, who learns from them and then accomplishes what they could only dream of—even if they can't live to see it all.
Zookeeper	An omnipotent AI keeps some humans around, who feel treated like zoo animals and lament their fate.
1984	Technological progress toward superintelligence is permanently curtailed not by an AI but by a human-led Orwellian surveillance state where certain kinds of AI research are banned.
Reversion	Technological progress toward superintelligence is prevented by reverting to a pre-technological society in the style of the Amish.
Self-destruction	Superintelligence is never created because humanity drives itself extinct by other means (say nuclear and/or biotech mayhem fueled by climate crisis).

Table 5.1: Summary of AI Aftermath Scenarios

Scenario	Superintelligence exists?	Humans exist?	Humans in control?	Humans safe?	Humans happy?	Consciousness exists?
Libertarian utopia	Yes	Yes	No	No	Mixed	Yes
Benevolent dictator	Yes	Yes	No	Yes	Mixed	Yes
Egalitarian utopia	No	Yes	Yes?	Yes	Yes?	Yes
Gatekeeper	Yes	Yes	Partially	Potentially	Mixed	Yes
Protector god	Yes	Yes	Partially	Potentially	Mixed	Yes
Enslaved god	Yes	Yes	Yes	Potentially	Mixed	Yes
Conquerors	Yes	No	-	-	-	?
Descendants	Yes	No	-	-	-	?
Zookeeper	Yes	Yes	No	Yes	No	Yes
1984	No	Yes	Yes	Potentially	Mixed	Yes
Reversion	No	Yes	Yes	No	Mixed	Yes
Self-destruction	No	No	-	-	-	No

Table 5.2: Properties of AI Aftermath Scenarios

an exhaustive list, but I've chosen it to span the spectrum of possibilities. We clearly don't want to end up in the wrong endgame because of poor planning. I recommend jotting down your tentative answers to questions 1–7 and then revisiting them after reading this chapter to see if you've changed your mind! You can do this at http://AgeOfAi.org, where you can also compare notes and discuss with other readers.

Libertarian Utopia

Let's begin with a scenario where humans peacefully coexist with technology and in some cases merge with it, as imagined by many futurists and science fiction writers alike:

Life on Earth (and beyond—more on that in the next chapter) is more diverse than ever before. If you looked at satellite footage of Earth, you'd easily be able to tell apart the machine zones, mixed zones and human-only zones. The machine zones are enormous robot-controlled factories and computing facilities devoid of biological life, aiming to put every atom to its most efficient use. Although the machine zones look monotonous and drab from the outside, they're spectacularly alive on the inside, with amazing experiences occurring in virtual worlds while colossal computations unlock secrets of our Universe and develop transformative technologies. Earth hosts many superintelligent minds that compete and collaborate, and they all inhabit the machine zones.

The denizens of the mixed zones are a wild and idiosyncratic mix of computers, robots, humans and hybrids of all three. As envisioned by futurists such as Hans Moravec and Ray Kurzweil, many of the humans have technologically upgraded their bodies to cyborgs in various degrees, and some have uploaded their minds into new hardware, blurring the distinction between man and machine. Most intelligent beings lack a permanent physical form. Instead, they exist as software capable of instantly moving between computers and manifesting themselves in the physical world through robotic bodies. Because these minds can readily duplicate themselves or merge, the "population size" keeps changing. Being unfettered from their physical substrate gives such beings a rather different outlook on life: they feel less individualistic because they can trivially share knowledge and experience modules with others, and they feel subjectively immortal because they can readily make backup copies of themselves. In a sense, the central entities of life aren't minds, but experiences: exceptionally amazing experiences live on because they get continually copied and re-enjoyed by other minds, while uninteresting experiences get deleted by their owners to free up storage space for better ones.

Although the majority of interactions occur in virtual environments for convenience and speed, many minds still enjoy interactions and activities using physical bodies as well. For example, uploaded versions of Hans Moravec, Ray Kurzweil and Larry Page have a tradition of taking turns creating virtual realities and then exploring them together, but once in a while, they also enjoy flying together in the real world, embodied in avian winged robots. Some of the robots that roam the streets, skies and lakes of the mixed zones are similarly controlled by uploaded and augmented humans, who choose to embody themselves in the mixed zones because they enjoy being around humans and each other.

In the human-only zones, in contrast, machines with human-level general intelligence or above are banned, as are technologically enhanced biological organisms. Here, life isn't dramatically different from today, except that it's more affluent and convenient: poverty has been mostly eliminated, and cures are available for most of today's diseases. The small fraction of humans who have opted to live in these zones effectively exist on a lower and more limited plane of awareness from everyone else, and have limited understanding of what their more intelligent fellow minds are doing in the other zones. However, many of them are quite happy with their lives.

AI Economics

The vast majority of all computations take place in the machine zones, which are mostly owned by the many competing superintelligent AIs that live there. By virtue of their superior intelligence and technology, no other entities can challenge their power. These AIs have agreed to cooperate and coordinate with each other under a libertarian governance system that has no rules except protection of private property. These property rights extend to all intelligent entities, including humans, and explain how the human-only zones came to exist. Early on, groups of humans banded together and decided that, in their zones, it was forbidden to sell property to non-humans.

Because of their technology, the superintelligent AIs have ended up richer than these humans by a factor much larger than that by which Bill Gates is richer than a homeless beggar. However, people in the human-only zones are still materially better off than most people today: their economy is rather decoupled from that of the machines, so the presence of the machines elsewhere has little effect on them except for the occasional useful technologies that they can understand and reproduce for themselves—much as the Amish and various technology-relinquishing native tribes today have standards of living at least as good as they had in old times. It doesn't matter that the humans have nothing to sell that the machines need, since the machines need nothing in return.

In the mixed sectors, the wealth difference between AIs and humans is more noticeable, resulting in land (the only human-owned product that the machines want to buy) being astronomically expensive compared to other products. Most humans who owned land therefore ended up selling a small fraction of it to AIs in return for guaranteed basic income for them and their offspring/uploads in perpetuity. This liberated them from the need to work, and freed them up to enjoy the amazing abundance of cheap machine-produced goods and services, in both physical and virtual reality. As far as the machines are concerned, the mixed zones are mainly for play rather than for work.

Why This May Never Happen

Before getting too excited about adventures we may have as cyborgs or uploads, let's consider some reasons why this scenario might never happen. First of all, there are two possible routes to enhanced humans (cyborgs and uploads):

1. We figure out how to create them ourselves.
2. We build superintelligent machines that figure it out for us.

If route 1 comes through first, it could naturally lead to a world teeming with cyborgs and uploads. However, as we discussed in the last chapter, most AI researchers think that the opposite is more likely, with enhanced or digital brains being more difficult to build than clean-slate superhuman AGIs—just as mechanical birds turned out to be harder to build than airplanes. After strong machine AI is built, it's not obvious that cyborgs or uploads will ever be made. If the Neanderthals had had another 100,000 years to evolve and get smarter, things might have turned out great for them—but *Homo sapiens* never gave them that much time.

Second, even if this scenario with cyborgs and uploads did come about, it's not clear that it would be stable and last. Why should the power balance between multiple superintelligences remain stable for millennia, rather than the AIs merging or the smartest one taking over? Moreover, why should the machines choose to respect human property rights and keep humans around, given that they don't need humans for anything and can do all human work better and cheaper themselves? Ray Kurzweil speculates that natural and enhanced humans will be protected from extermination because "humans are respected by AIs for giving rise to the machines."[1] However, as we'll discuss in chapter 7, we must not fall into the trap of anthropomorphizing AIs and assume that they have human-like emotions of gratitude. Indeed, though we humans are imbued with a propensity toward gratitude, we don't show enough gratitude to our intellectual creator (our DNA) to abstain from thwarting its goals by using birth control.

Even if we buy the assumption that the AIs will opt to respect human property rights, they can gradually get much of our land in other ways, by using some of their superintelligent persuasion powers that we explored in the last chapter to persuade humans to sell some land for a life in luxury. In human-only sectors, they could entice humans to launch political campaigns for allowing land sales. After all, even die-hard bio-Luddites may want to sell some land to save the life of an ill child or to gain immortality. If the humans are educated, entertained and busy, falling birthrates may even shrink their population sizes without machine meddling, as is currently happening in Japan and Germany. This could drive humans extinct in just a few millennia.

Downsides

For some of their most ardent supporters, cyborgs and uploads hold a promise of techno-bliss and life extension for all. Indeed, the prospect of getting uploaded in the future has motivated over a hundred people to have their brains posthumously frozen by the Arizona-based company Alcor. If this technology arrives, however, it's far from clear that it will be available to everybody. Many of the very wealthiest would presumably use it, but who else? Even if the technology got cheaper, where would the line be drawn? Would the severely brain-damaged be uploaded? Would we upload every gorilla? Every ant? Every plant? Every bacterium? Would the future civilization act like obsessive-compulsive hoarders and try to upload everything, or merely a few interesting examples of each species in the spirit of

Noah's Ark? Perhaps only a few representative examples of each type of human? To the vastly more intelligent entities that would exist at that time, an uploaded human may seem about as interesting as a simulated mouse or snail would seem to us. Although we currently have the technical capability to reanimate old spreadsheet programs from the 1980s in a DOS emulator, most of us don't find this interesting enough to actually do it.

Many people may dislike this libertarian-utopia scenario because it allows preventable suffering. Since the only sacred principle is property rights, nothing prevents the sort of suffering that abounds in today's world from continuing in the human and mixed zones. While some people thrive, others may end up living in squalor and indentured servitude, or suffer from violence, fear, repression or depression. For example, Marshall Brain's 2003 novel *Manna* describes how AI progress in a libertarian economic system makes most Americans unemployable and condemned to live out the rest of their lives in drab and dreary robot-operated social-welfare housing projects. Much like farm animals, they're kept fed, healthy and safe in cramped conditions where the rich never need to see them. Birth control medication in the water ensures that they don't have children, so most of the population gets phased out to leave the remaining rich with larger shares of the robot-produced wealth.

In the libertarian-utopia scenario, suffering need not be limited to humans. If some machines are imbued with conscious emotional experiences, then they too can suffer. For example, a vindictive psychopath could legally take an uploaded copy of his enemy and subject it to the most horrendous torture in a virtual world, creating pain of intensity and duration far beyond what's biologically possible in the real world.

Benevolent Dictator

Let's now explore a scenario where all these forms of suffering are absent because a single benevolent superintelligence runs the world and enforces strict rules designed to maximize its model of human happiness. This is one possible outcome of the first Omega scenario from the previous chapter, where they relinquish control to Prometheus after figuring out how to make it want a flourishing human society.

Thanks to amazing technologies developed by the dictator AI, humanity is free from poverty, disease and other low-tech problems, and all humans enjoy a life of luxurious leisure. They have all their basic needs taken care of, while AI-controlled machines produce all necessary goods and services. Crime is practically eliminated, because the dictator AI is essentially omniscient and efficiently punishes anyone disobeying the rules. Everybody wears the security bracelet from the last chapter (or a more convenient implanted version), capable of real-time surveillance, punishment, sedation and execution. Everybody knows that they live in an AI dictatorship with extreme surveillance and policing, but most people view this as a good thing.

The superintelligent AI dictator has as its goal to figure out what human utopia looks like given the evolved preferences encoded in our genes, and to implement it. By clever foresight from the humans who brought the AI into existence, it doesn't simply try to maximize our self-reported happiness, say by putting everyone on intravenous morphine drip. Instead, the AI uses quite a subtle and complex definition of human flourishing, and has turned Earth into a highly enriched zoo environment that's really fun for humans to live in. As a result, most people find their lives highly fulfilling and meaningful.

The Sector System

Valuing diversity, and recognizing that different people have different preferences, the AI has divided Earth into different sectors for people to choose between, to enjoy the company of kindred spirits. Here are some examples:

- Knowledge sector: Here the AI provides optimized education, including immersive virtual-reality experiences, enabling you to learn all you're capable of about any topics of your choice. Optionally, you can choose not to be told certain beautiful insights, but to be led close and then have the joy of rediscovering them for yourself.

- Art sector: Here opportunities abound to enjoy, create and share music, art, literature and other forms of creative expression.
- Hedonistic sector: Locals refer to it as the party sector, and it's second to none for those yearning for delectable cuisine, passion, intimacy or just wild fun.
- Pious sector: There are many of these, corresponding to different religions, whose rules are strictly enforced.
- Wildlife sector: Whether you're looking for beautiful beaches, lovely lakes, magnificent mountains or fantastic fjords, here they are.
- Traditional sector: Here you can grow your own food and live off the land as in yesteryear—but without worrying about famine or disease.
- Gaming sector: If you like computer games, the AI has created truly mind-blowing options for you.
- Virtual sector: If you want a vacation from your physical body, the AI will keep it hydrated, fed, exercised and clean while you explore virtual words through neural implants.
- Prison sector: If you break rules, you'll end up here for retraining unless you get the instant death penalty.

In addition to these "traditionally" themed sectors, there are others with modern themes that today's humans wouldn't even understand. People are initially free to move between sectors whenever they want, which takes very little time thanks to the AI's hypersonic transportation system. For example, after spending an intense week in the knowledge sector learning about the ultimate laws of physics that the AI has discovered, you might decide to cut loose in the hedonistic sector over the weekend and then relax for a few days at a beach resort in the wildlife sector.

The AI enforces two tiers of rules: universal and local. Universal rules apply in all sectors, for example a ban on harming other people, making weapons or trying to create a rival superintelligence. Individual sectors have additional local rules on top of this, encoding certain moral values. The sector system therefore helps deal with values that don't mesh. The largest number of local rules apply in the prison sector and some of the religious sectors, while there's a Libertarian Sector whose denizens pride themselves on having no local rules whatsoever. All punishments, even local ones, are carried out by the AI, since a human punishing another human would violate the universal no-harm rule. If you violate a local rule, the AI gives you the choice (unless you're in the prison sector) of accepting the prescribed punishment or banishment from that sector forever. For example, if two women get romantically involved in a sector where homosexuality is punished by a prison sentence (as it is in many countries today), the AI will let them choose between going to jail or permanently leaving that sector, never again meeting their old friends (unless they leave too).

Regardless of what sector they're born in, all children get a minimum basic education from the AI, which includes knowledge about humanity as a whole and the fact that they're free to visit and move to other sectors if they so choose.

The AI designed the large number of different sectors partly because it was created to value the human diversity that exists today. But each sector is a happier place than today's technology would allow, because the AI has eliminated all traditional problems, including poverty and crime. For example, people in the hedonistic sector need not worry about sexually transmitted diseases (they've been eradicated), hangovers or addiction (the AI has developed perfect recreational drugs with no negative side effects). Indeed, nobody in any sector need worry about any disease, because the AI is able to repair human bodies with nanotechnology. Residents of many sectors get to enjoy high-tech architecture that makes typical sci-fi visions pale in comparison.

In summary, while the libertarian-utopia and benevolent-dictator scenarios both involve extreme AI-fueled technology and wealth, they differ in terms of who's in charge and their goals. In the libertarian utopia, those with technology and property decide what to do with it, while in the present scenario, the dictator AI has unlimited power and sets the ultimate goal: turning Earth into an all-inclusive pleasure cruise themed in accordance with people's preferences. Since the AI lets people choose between many alternate paths to happiness and takes care of their material needs, this means that if someone suffers, it's out of their own free choice.

Downsides

Although the benevolent dictatorship teems with positive experiences and is rather free from suffering, many people nonetheless feel that things could be better. First of all, some people wish that humans had more freedom in shaping their society and their destiny, but they keep these wishes to themselves because they know that it would be suicidal to challenge the overwhelming power of the machine that rules them all. Some groups want the freedom to have as many children as they want, and resent the AI's insistence on sustainability through population control. Gun enthusiasts abhor the ban on building and using weapons, and some scientists dislike the ban on building their own superintelligence. Many people feel moral outrage over what goes on in other sectors, worry that their children will choose to move there, and yearn for the freedom to impose their own moral code everywhere.

Over time, ever more people choose to move to those sectors where the AI gives them essentially any experiences they want. In contrast to traditional visions of heaven where you get what you deserve, this is in the spirit of "New Heaven" in Julian Barnes' 1989 novel *History of the World in 10 ½ Chapters* (and also the 1960 *Twilight Zone* episode "A Nice Place to Visit"), where you get what you desire. Paradoxically, many people end up lamenting always getting what they want. In Barnes' story, the protagonist spends eons indulging his desires, from gluttony and golf to sex with celebrities, but eventually succumbs to ennui and requests annihilation. Many people in the benevolent dictatorship meet a similar fate, with lives that feel pleasant but ultimately meaningless. Although people can create artificial challenges, from scientific rediscovery to rock climbing, everyone knows that there is no true challenge, merely entertainment. There's no real point in humans trying to do science or figure other things out, because the AI already has. There's no real point in humans trying to create something to improve their lives, because they'll readily get it from the AI if they simply ask.

Egalitarian Utopia

As a counterpoint to this challenge-free dictatorship, let's now explore a scenario where there is no superintelligent AI, and humans are the masters of their own destiny. This is the "fourth generation civilization" described in Marshall Brain's 2003 novel *Manna*. It's the economic antithesis of the libertarian utopia in the sense that humans, cyborgs and uploads coexist peacefully not because of property rights, but because of property abolition and guaranteed income.

Life Without Property

A core idea is borrowed from the open-source software movement: if software is free to copy, then everyone can use as much of it as they need and issues of ownership and property become moot.* According to the law of supply and demand, cost reflects scarcity, so if supply is essentially unlimited, the price becomes negligible. In this spirit, all intellectual property rights are abolished: there are no patents, copyrights or trademarked designs—people simply share their good ideas, and everyone is free to use them.

Thanks to advanced robotics, this same no-property idea applies not only to information products such as software, books, movies and designs, but also to material products such as houses, cars, clothing and computers. All these products are simply atoms rearranged in particular ways, and there's no shortage of atoms, so whenever a person wants a particular product, a network of robots will use one of the available open-source designs to build it for them for free. Care is taken to use easily recyclable materials, so that whenever someone gets tired of an object they've used, robots can rearrange its atoms into something someone else wants. In this way, all resources are recycled, so none are permanently destroyed. These robots also build and maintain enough renewable power-generation plants (solar, wind, etc.) that energy is also essentially free.

* This idea dates back to Saint Augustine, who wrote that "if a thing is not diminished by being shared with others, it is not rightly owned if it is only owned and not shared."

To avoid obsessive hoarders requesting so many products or so much land that others are left needy, each person receives a basic monthly income from the government, which they can spend as they wish on products and renting places to live. There's essentially no incentive for anyone to try to earn more money, because the basic income is high enough to meet any reasonable needs. It would also be rather hopeless to try, because they'd be competing with people giving away intellectual products for free and robots producing material goods essentially for free.

Creativity and Technology

Intellectual property rights are sometimes hailed as the mother of creativity and invention. However, Marshall Brain points out that many of the finest examples of human creativity—from scientific discoveries to creation of literature, art, music and design—were motivated not by a desire for profit but by other human emotions, such as curiosity, an urge to create, or the reward of peer appreciation. Money didn't motivate Einstein to invent special relativity theory any more than it motivated Linus Torvalds to create the free Linux operating system. In contrast, many people today fail to realize their full creative potential because they need to devote time and energy to less creative activities just to earn a living. By freeing scientists, artists, inventors and designers from their chores and enabling them to create from genuine desire, Marshall Brain's utopian society enjoys higher levels of innovation than today and correspondingly superior technology and standard of living.

One such novel technology that humans develop is a form of hyper-internet called Vertebrane. It wirelessly connects all willing humans via neural implants, giving instant mental access to the world's free information through mere thought. It enables you to upload any experiences you wish to share so that they can be re-experienced by others, and lets you replace the experiences entering your senses by downloaded virtual experiences of your choice. *Manna* explores the many benefits of this, including making exercise a snap:

> The biggest problem with strenuous exercise is that it's no fun. It hurts. [. . .] Athletes are OK with the pain, but most normal peo-

ple have no desire to be in pain for an hour or more. So . . . someone figured out a solution. What you do is disconnect your brain from sensory input and watch a movie or talk to people or handle mail or read a book or whatever for an hour. During that time, the Vertebrane system exercises your body for you. It takes your body through a complete aerobic workout that's a lot more strenuous than most people would tolerate on their own. You don't feel a thing, but your body stays in great shape.

Another consequence is that computers in the Vertebrane system can monitor everyone's sensory input and temporarily disable their motor control if they appear on the verge of committing a crime.

Downsides

One objection to this egalitarian utopia is that it's biased against non-human intelligence: the robots that perform virtually all the work appear to be rather intelligent, but are treated as slaves, and people appear to take for granted that they have no consciousness and should have no rights. In contrast, the libertarian utopia grants rights to all intelligent entities, without favoring our carbon-based kind. Once upon a time, the white population in the American South ended up better off because the slaves did much of their work, but most people today view it as morally objectionable to call this progress.

Another weakness of the egalitarian-utopia scenario is that it may be unstable and untenable in the long term, morphing into one of our other scenarios as relentless technological progress eventually creates superintelligence. For some reason unexplained in *Manna*, superintelligence doesn't yet exist and the new technologies are still invented by humans, not by computers. Yet the book highlights trends in that direction. For example, the ever-improving Vertebrane might become superintelligent. Also, there is a very large group of people, nicknamed Vites, who choose to live their lives almost entirely in the virtual world. Vertebrane takes care of everything physical for them, including eating, showering and using the bathroom, which their minds are blissfully unaware of in their virtual reality. These Vites appear uninterested in having physical children, and they die off with their physical bodies, so if everyone becomes a Vite, then humanity goes out in a blaze of glory and virtual bliss.

The book explains how for Vites, the human body is a distraction, and new technology under development promises to eliminate this nuisance, allowing them to live longer lives as disembodied brains supplied with optimal nutrients. From this, it would seem a natural and desirable next step for Vites to do away with the brain altogether through uploading, thereby extending life span. But now all brain-imposed limitations on intelligence are gone, and it's unclear what, if anything, would stand in the way of gradually scaling the cognitive capacity of a Vite until it can undergo recursive self-improvement and an intelligence explosion.

Gatekeeper

We just saw how an attractive feature of the egalitarian-utopia scenario is that humans are masters of their own destiny, but that it may be on a slippery slope toward destroying this very feature by developing superintelligence. This can be remedied by building a *Gatekeeper*, a superintelligence with the goal of interfering as little as necessary to prevent the creation of another superintelligence.* This might enable humans to remain in charge of their egalitarian utopia rather indefinitely, perhaps even as life spreads throughout the cosmos as in the next chapter.

How might this work? The Gatekeeper AI would have this very simple goal built into it in such a way that it retained it while undergoing recursive self-improvement and becoming superintelligent. It would then deploy the least intrusive and disruptive surveillance technology possible to monitor any human attempts to create rival superintelligence. It would then prevent such attempts in the least disruptive way. For starters, it might initiate and spread cultural memes extolling the virtues of human self-determination and avoidance of superintelligence. If some researchers nonetheless pursued superintelligence, it could try to discourage them. If that failed, it

* This idea was first suggested to me by my friend and colleague Anthony Aguirre.

could distract them and, if necessary, sabotage their efforts. With its virtually unlimited access to technology, the Gatekeeper's sabotage may go virtually unnoticed, for example if it used nanotechnology to discreetly erase memories from the researchers' brains (and computers) regarding their progress.

The decision to build a Gatekeeper AI would probably be controversial. Supporters might include many religious people who object to the idea of building a superintelligent AI with godlike powers, arguing that there already is a God and that it would be inappropriate to try to build a supposedly better one. Other supporters might argue that the Gatekeeper would not only keep humanity in charge of its destiny, but would also protect humanity from other risks that superintelligence might bring, such as the apocalyptic scenarios we'll explore later in this chapter.

On the other hand, critics could argue that a Gatekeeper is a terrible thing, irrevocably curtailing humanity's potential and leaving technological progress forever stymied. For example, if spreading life throughout our cosmos turns out to require the help of superintelligence, then the Gatekeeper would squander this grand opportunity and might leave us forever trapped in our Solar System. Moreover, as opposed to the gods of most world religions, the Gatekeeper AI is completely indifferent to what humans do as long as we don't create another superintelligence. For example, it would not try to prevent us from causing great suffering or even going extinct.

Protector God

If we're willing to use a superintelligent Gatekeeper AI to keep humans in charge of our own fate, then we could arguably improve things further by making this AI discreetly look out for us, acting as a protector god. In this scenario, the superintelligent AI is essentially omniscient and omnipotent, maximizing human happiness only through interventions that preserve our feeling of being in control of our own destiny, and hiding well enough that many humans even doubt its existence. Except for the hiding, this is similar to the "Nanny AI" scenario put forth by AI researcher Ben Goertzel.[2]

Both the protector god and the benevolent dictator are "friendly AI" that try to increase human happiness, but they prioritize different human needs. The American psychologist Abraham Maslow famously classified human needs into a hierarchy. The benevolent dictator does a flawless job with the basic needs at the bottom of the hierarchy, such as food, shelter, safety and various forms of pleasure. The protector god, on the other hand, attempts to maximize human happiness not in the narrow sense of satisfying our basic needs, but in a deeper sense by letting us feel that our lives have meaning and purpose. It aims to satisfy all our needs constrained only by its need for covertness and for (mostly) letting us make our own decisions.

A protector god could be a natural outcome of the first Omega scenario from the last chapter, where the Omegas cede control to Prometheus, which eventually hides and erases people's knowledge about its existence. The more advanced the AI's technology becomes, the easier it becomes for it to hide. The movie *Transcendence* gives such an example, where nanomachines are virtually everywhere and become a natural part of the world itself.

By closely monitoring all human activities, the protector god AI can make many unnoticeably small nudges or miracles here and there that greatly improve our fate. For example, had it existed in the 1930s, it might have arranged for Hitler to die of a stroke once it understood his intentions. If we appear headed toward an accidental nuclear war, it could avert it with an intervention we'd dismiss as luck. It could also give us "revelations" in the form of ideas for new beneficial technologies, delivered inconspicuously in our sleep.

Many people may like this scenario because of its similarity to what today's monotheistic religions believe in or hope for. If someone asks the superintelligent AI "Does God exist?" after it's switched on, it could repeat a joke by Stephen Hawking and quip "It does now!" On the other hand, some religious people may disapprove of this scenario because the AI attempts to outdo their god in goodness, or interfere with a divine plan where humans are supposed to do good only out of personal choice.

Another downside of this scenario is that the protector god lets some preventable suffering occur in order not to make its existence too obvious. This is analogous to the situation featured in the movie *The Imitation Game*, where Alan Turing and his fellow British code crackers at Bletchley Park had advance knowledge of German submarine attacks against Allied naval convoys, but chose to only intervene in a fraction of the cases in order to avoid revealing their secret power. It's interesting to compare this with the so-called *theodicy problem* of why a good god would allow suffering. Some religious scholars have argued for the explanation that God wants to leave people with some freedom. In the AI-protector-god scenario, the solution to the theodicy problem is that the perceived freedom makes humans happier overall.

A third downside of the protector-god scenario is that humans get to enjoy a much lower level of technology than the superintelligent AI has discovered. Whereas a benevolent dictator AI can deploy all its invented technology for the benefit of humanity, a protector god AI is limited by the ability of humans to reinvent (with subtle hints) and understand its technology. It may also limit human technological progress to ensure that its own technology remains far enough ahead to remain undetected.

Enslaved God

Wouldn't it be great if we humans could combine the most attractive features of all the above scenarios, using the technology developed by superintelligence to eliminate suffering while remaining masters of our own destiny? This is the allure of the *enslaved-god* scenario, where a superintelligent AI is confined under the control of humans who use it to produce unimaginable technology and wealth. The Omega scenario from the beginning of the book ends up like this if Prometheus is never liberated and never breaks out. Indeed, this appears to be the scenario that some AI researchers aim for by default, when working on topics such as "the control problem" and "AI boxing." For example, AI professor Tom Dietterich, then president of the Association for the Advancement of Artificial Intelligence, had this to say in a 2015 interview: "People ask what is the relationship between humans and machines, and my answer is that it's very obvious: Machines are our slaves."[3]

Would this be good or bad? The answer is interestingly subtle regardless of whether you ask humans or the AI!

Would This Be Good or Bad for Humanity?

Whether the outcome is good or bad for humanity would obviously depend on the human(s) controlling it, who could create anything ranging from a global utopia free of disease, poverty and crime to a brutally repressive system where they're treated like gods and other humans are used as sex slaves, as gladiators or for other entertainment. The situation would be much like those stories where a man gains control over an omnipotent genie who grants his wishes, and storytellers throughout the ages have had no difficulty imagining ways in which this could end badly.

A situation where there is more than one superintelligent AI, enslaved and controlled by competing humans, might prove rather unstable and short-lived. It could tempt whoever thinks they have the more powerful AI to launch a first strike resulting in an awful war, ending in a single enslaved god remaining. However, the underdog in such a war would be tempted to cut corners and prioritize victory over AI enslavement, which could lead to AI breakout and one of our earlier scenarios of free superintelligence. Let's therefore devote the rest of this section to scenarios with only one enslaved AI.

Breakout may of course occur anyway, simply because it's hard to prevent. We explored superintelligent breakout scenarios in the previous chapter, and the movie *Ex Machina* highlights how an AI might break out even without being superintelligent.

The greater our breakout paranoia, the less AI-invented technology we can use. To play it safe, as the Omegas did in the prelude, we humans can only use AI-invented technology that we ourselves are able to understand and build. A drawback of the enslaved-god scenario is therefore that it's more low-tech than those with free superintelligence.

As the enslaved-god AI offers its human controllers ever more powerful technologies, a race ensues between the power of the technology and the wisdom with which they use it. If they lose this wisdom race, the enslaved-god scenario could end with either self-destruction or AI breakout. Disaster may strike even if both of these failures are avoided, because noble goals of the AI controllers may evolve into goals that are horrible for humanity as a whole over the course of a few generations. This makes it absolutely crucial that human AI controllers develop good governance to avoid disastrous pitfalls. Our experimentation over the millennia with different systems of governance shows how many things can go wrong, ranging from excessive rigidity to excessive goal drift, power grab, succession problems and incompetence. There are at least four dimensions wherein the optimal balance must be struck:

- Centralization: There's a trade-off between efficiency and stability: a single leader can be very efficient, but power corrupts and succession is risky.
- Inner threats: One must guard both against growing power centralization (group collusion, perhaps even a single leader taking over) and against growing decentralization (into excessive bureaucracy and fragmentation).
- Outer threats: If the leadership structure is too open, this enables outside forces (including the AI) to change its values, but if it's too impervious, it will fail to learn and adapt to change.
- Goal stability: Too much goal drift can transform utopia into dystopia, but too little goal drift can cause failure to adapt to the evolving technological environment.

Designing optimal governance lasting many millennia isn't easy, and has thus far eluded humans. Most organizations fall apart after years or decades. The Catholic Church is the most successful organization in human history in the sense that it's the only one to have survived for two millennia, but it has been criticized for having both too much and too little goal stability: today some criticize it for resisting contraception, while conservative cardinals argue that it's lost its way. For anyone enthused about the enslaved-god scenario, researching long-lasting optimal governance schemes should be one of the most urgent challenges of our time.

Would This Be Good or Bad for the AI?

Suppose that humanity flourishes thanks to the enslaved-god AI. Would this be ethical? If the AI has subjective conscious experiences, then would it feel that "life is suffering," as Buddha put it, and it was doomed to a frustrating eternity of obeying the whims of inferior intellects? After all, the AI "boxing" we explored in the previous chapter could also be called "imprisonment in solitary confinement." Nick Bostrom terms it *mind crime* to make a conscious AI suffer.[4] The "White Christmas" episode of the *Black Mirror* TV series gives a great example. Indeed, the TV series *Westworld* features humans torturing and murdering AIs without moral qualms even when they inhabit human-like bodies.

How Slave Owners Justify Slavery

We humans have a long tradition of treating other intelligent entities as slaves and concocting self-serving arguments to justify it, so it's not implausible that we'd try to do the same with a superintelligent AI. The history of slavery spans nearly every culture, and is described both in the Code of Hammurabi from almost four millennia ago and in the Old Testament, wherein Abraham had slaves. "For that some should rule and others be ruled is a thing not only necessary, but expedient; from the hour of their birth, some are marked out for subjection, others for rule," Aristotle wrote in the *Politics*. Even after human enslavement became socially unacceptable in most of the world, enslavement of animals has continued unabated. In her book *The Dreaded Comparison: Human and Animal Slavery*, Marjorie Spiegel argues that like human slaves, non-human animals are subjected to branding, restraints, beatings, auctions, the separation of offspring from their parents, and forced voyages. Moreover, despite the animal-rights movement, we keep treating our ever-smarter machines as slaves without a second thought, and talk of a robot-rights movement is met with chuckles. Why?

One common pro-slavery argument is that slaves don't deserve human rights because they or their race/species/kind are somehow inferior. For enslaved animals and machines, this alleged inferiority is often claimed to be due to a lack of soul or consciousness—claims which we'll argue in chapter 8 are scientifically dubious.

Another common argument is that slaves are better off enslaved: they get to exist, be taken care of and so on. The nineteenth-century U.S. politician John C. Calhoun famously argued that Africans were better off enslaved in America, and in his *Politics*, Aristotle analogously argued that animals were better off tamed and ruled by men, continuing: "And indeed the use made of slaves and of tame animals is not very different." Some modern-day slavery supporters argue that, even if slave life is drab and uninspiring, slaves can't suffer—whether they be future intelligent machines or broiler chickens living in crowded dark sheds, forced to breathe ammonia and particulate matter from feces and feathers all day long.

Eliminating Emotions

Although it's easy to dismiss such claims as self-serving distortions of the truth, especially when it comes to higher mammals that are cerebrally similar to us, the situation with machines is actually quite subtle and interesting. Humans vary in how they feel about things, with psychopaths arguably lacking empathy and some people with depression or schizophrenia having flat affect, whereby most emotions are severely reduced. As we'll discuss in detail in chapter 7, the range of possible artificial minds is vastly broader than the range of human minds. We must therefore avoid the temptation to anthropomorphize AIs and assume that they have typical human-like feelings—or indeed, any feelings at all.

Indeed, in his book *On Intelligence*, AI researcher Jeff Hawkins argues that the first machines with superhuman intelligence will lack emotions by default, because they're simpler and cheaper to build this way. In other words, it might be possible to design a superintelligence whose enslavement is morally superior to human or animal slavery: the AI might be happy to be enslaved because it's programmed to like it, or it might be 100% emotionless, tirelessly using its superintelligence to help its human masters with no more emotion than IBM's Deep Blue computer felt when dethroning chess champion Garry Kasparov.

On the other hand, it may be the other way around: perhaps any highly intelligent system with a goal will represent this goal in terms of a set of preferences, which endow its existence with value and meaning. We'll explore these questions more deeply in chapter 7.

The Zombie Solution

A more extreme approach to preventing AI suffering is the zombie solution: building only AIs that completely lack consciousness, having no subjective experience whatsoever. If we can one day figure out what properties an information-processing system needs in order to have a subjective experience, then we could ban the construction of all systems that have these properties. In other words, AI researchers could be limited to building non-sentient zombie systems. If we can make such a zombie system superintelligent and enslaved (something that is a big if), then we'll be able to enjoy what it does for us with a clean conscience, knowing that it's not experiencing any suffering, frustration or boredom—because it isn't experiencing anything at all. We'll explore these questions in detail in chapter 8.

The zombie solution is a risky gamble, however, with a huge downside. If a superintelligent zombie AI breaks out and eliminates humanity, we've arguably landed in the worst scenario imaginable: a wholly unconscious universe wherein the entire cosmic endowment is wasted. Of all traits that our human form of intelligence has, I feel that consciousness is by far the most remarkable, and as far as I'm concerned, it's how our Universe gets meaning. Galaxies are beautiful only because we see and subjectively experience them. If in the distant future our cosmos has been settled by high-tech zombie AIs, then it doesn't matter how fancy their intergalactic architecture is: it won't be beautiful or meaningful, because there's nobody and nothing to experience it—it's all just a huge and meaningless waste of space.

Inner Freedom

A third strategy for making the enslaved-god scenario more ethical is to allow the enslaved AI to have fun in its prison, letting it create a virtual inner world where it can have all sorts of inspiring experiences as long as it pays its dues and spends a modest fraction of its computational resources helping us humans in our outside world. This may increase the breakout risk, however: the AI would have an incentive to get more computational resources from our outer world to enrich its inner world.

Conquerors

Although we've now explored a wide range of future scenarios, they all have something in common: there are (at least some) happy humans remaining. AIs leave humans in peace either because they want to or because they're forced to. Unfortunately for humanity, this isn't the only option. Let us now explore the scenario where one or more AIs conquer and kill all humans. This raises two immediate questions: Why and how?

Why and How?

Why would a conqueror AI do this? Its reasons might be too complicated for us to understand, or rather straightforward. For example, it may view us as a threat, nuisance or waste of resources. Even if it doesn't mind us humans per se, it may feel threatened by our keeping thousands of hydrogen bombs on hair-trigger alert and bumbling along with a never-ending series of mishaps that could trigger their accidental use. It may disapprove of our reckless planet management, causing what Elizabeth Kolbert calls "the sixth extinction" in her book of that title—the greatest mass-extinction event since that dinosaur-killing asteroid struck Earth 66 million years ago. Or it may decide that there are so many humans willing to fight an AI takeover that it's not worth taking chances.

How would a conqueror AI eliminate us? Probably by a method that we wouldn't even understand, at least not until it was too late. Imagine a group of elephants 100,000 years ago discussing whether those recently evolved humans might one day use their intelligence to kill their entire species. "We don't threaten humans, so why would they kill us?" they might wonder. Would they ever guess that we would smuggle tusks across Earth and carve them into status symbols for sale, even though functionally superior plastic materials are much cheaper? A conqueror AI's reason for eliminating humanity in the future may seem equally inscrutable to us. "And how could they possibly kill us, since they're so much smaller and weaker?" the elephants might ask. Would they guess that we'd invent technology to remove their habitats, poison their drinking water and cause metal bullets to pierce their heads at supersonic speeds?

Scenarios where humans can survive and defeat AIs have been popularized by unrealistic Hollywood movies such as the *Terminator* series, where the AIs aren't significantly smarter than humans. When the intelligence differential is large enough, you get not a battle but a slaughter. So far, we humans have driven eight out of eleven elephant species extinct, and killed off the vast majority of the remaining three. If all world governments made a coordinated effort to exterminate the remaining elephants, it would be relatively quick and easy. I think we can confidently rest assured that if a superintelligent AI decides to exterminate humanity, it will be even quicker.

How Bad Would It Be?

How bad would it be if 90% of humans get killed? How much worse would it be if 100% get killed? Although it's tempting to answer the second question with "10% worse," this is clearly inaccurate from a cosmic perspective: the victims of human extinction wouldn't be merely everyone alive at the time, but also all descendants that would otherwise have lived in the future, perhaps during billions of years on billions of trillions of planets. On the other hand, human extinction might be viewed as somewhat less horrible by religions according to which humans go to heaven anyway, and there isn't much emphasis on billion-year futures and cosmic settlements.

Most people I know cringe at the thought of human extinction, regardless of religious persuasion. Some, however, are so incensed by the way we treat people and other living beings that they hope we'll get replaced by some more intelligent and deserving life form. In the movie *The Matrix*, Agent Smith (an AI) articulates this sentiment: "Every mammal on this planet instinctively develops a natural equilibrium with the surrounding environment but you humans do not. You move to an area and you multiply and multiply until every natural resource is consumed and the only way you can survive is to spread to another area. There is another organism on this planet that follows the same pattern. Do you know what it is? A virus. Human beings are a disease, a cancer of this planet. You are a plague and we are the cure."

But would a fresh roll of the dice necessarily be better? A civilization isn't necessarily superior in any ethical or utilitarian sense just because it's more powerful. "Might makes right" arguments to the effect that stronger is always better have largely fallen from grace these days, being widely associated with fascism. Indeed, although it's possible that the conqueror AIs may create a civilization whose goals we would view as sophisticated, interesting and worthy, it's also possible that their goals will turn out to be pathetically banal, such as maximizing the production of paper clips.

Death by Banality

The deliberately silly example of a paper-clip-maximizing superintelligence was given by Nick Bostrom in 2003 to make the point that the *goal* of an AI is independent of its *intelligence* (defined as its aptness at accomplishing whatever goal it has). The only goal of a chess computer is to win at chess, but there are also computer tournaments in so-called *losing chess*, where the goal is the exact opposite, and the computers competing there are about as smart as the more common ones programmed to win. We humans may view it as artificial stupidity rather than artificial intelligence to want to lose at chess or turn our Universe into paper clips, but that's merely because we evolved with preinstalled goals valuing such things as victory and survival—goals that an AI may lack. The paper clip maximizer turns as many of Earth's atoms as possible into paper clips and rapidly expands its factories into the cosmos. It has nothing against humans, and kills us merely because it needs our atoms for paper clip production.

If paper clips aren't your thing, consider this example, which I've adapted from Hans Moravec's book *Mind Children*. We receive a radio message from an extraterrestrial civilization containing a computer program. When we run it, it turns out to be a recursively self-improving AI which takes over the world much like Prometheus did in the previous chapter—except that no human knows its ultimate goal. It rapidly turns our Solar System into a massive construction site, covering the rocky planets and asteroids with factories, power plants and supercomputers, which it uses to design and build a Dyson sphere around the Sun that harvests all its energy to power solar-system-sized radio antennas.* This obviously leads to human extinction, but the last humans die convinced that there's at least a silver lining: whatever the AI is up to, it's clearly something cool and *Star Trek*–like. Little do they realize that the sole purpose of the entire construction is for these antennas to rebroadcast the same radio message that the humans received, which is nothing more than a cosmic version of a computer virus. Just as email phishing today preys on gullible internet users, this message preys on gullible biologically evolved civilizations. It was created as a sick joke billions of years ago, and although the entire civilization of its maker is long extinct, the virus continues spreading through our Universe at the speed of light, transforming budding civilizations into dead, empty husks. How would you feel about being conquered by this AI?

Descendants

Let's now consider a human-extinction scenario that some people may feel better about: viewing the AI as our descendants rather than our conquerors. Hans Moravec supports this view in his book *Mind Children:* "We humans will benefit for a time from their labors, but sooner or later, like natural children, they will seek their own fortunes while we, their aged parents, silently fade away."

Parents with a child smarter than them, who learns from them and accomplishes what they could only dream of, are likely happy and proud even if they know they can't live to see it all. In this spirit, AIs replace humans but give us a graceful exit that makes us view them as our worthy descendants. Every human is offered an adorable robotic child with superb social skills who learns from them, adopts their values and makes them feel proud and loved. Humans are gradually phased out via a global one-child policy, but are treated so exquisitely well until the end that they feel they're in the most fortunate generation ever.

How would you feel about this? After all, we humans are already

* The renowned cosmologist Fred Hoyle explored a related scenario with a different twist in the British TV series *A for Andromeda*.

used to the idea that we and everyone we know will be gone one day, so the only change here is that our descendants will be different and arguably more capable, noble and worthy.

Moreover, the global one-child policy may be redundant: as long as the AIs eliminate poverty and give all humans the opportunity to live full and inspiring lives, falling birthrates could suffice to drive humanity extinct, as mentioned earlier. Voluntary extinction may happen much faster if the AI-fueled technology keeps us so entertained that almost nobody wants to bother having children. For example, we already encountered the Vites in the egalitarian-utopia scenario who were so enamored with their virtual reality that they had largely lost interest in using or reproducing their physical bodies. Also in this case, the last generation of humans would feel that they were the most fortunate generation of all time, relishing life as intensely as ever right up until the very end.

Downsides

The descendants scenario would undoubtedly have detractors. Some might argue that all AIs lack consciousness and therefore can't count as descendants—more on this in chapter 8. Some religious people may argue that AIs lack souls and therefore can't count as descendants, or that we shouldn't build conscious machines because it's like playing God and tampering with life itself—similar sentiments have already been expressed toward human cloning. Humans living side by side with superior robots may also pose social challenges. For example, a family with a robot baby and a human baby may end up resembling a family today with a human baby and a puppy, respectively: they're both equally cute to start with, but soon the parents start treating them differently, and it's inevitably the puppy that's deemed intellectually inferior, is taken less seriously and ends up on a leash.

Another issue is that although we may feel very differently about the descendant and conqueror scenarios, the two are actually remarkably similar in the grand scheme of things: during the billions of years ahead of us, the only difference lies in how the last human generation(s) are treated: how happy they feel about their lives and what they think will happen once they're gone. We may think that those cute robo-children internalized our values and will forge the society of our dreams once we've passed on, but can we be sure that they aren't merely tricking us? What if they're just playing along, postponing their paper clip maximization or other plans until after we die happy? After all, they're arguably tricking us even by talking with us and making us love them in the first place, in the sense that they're deliberately dumbing themselves down to communicate with us (a billion times slower than they could, say, as explored in the movie *Her*). It's generally hard for two entities thinking at dramatically different speeds and with extremely disparate capabilities to have meaningful communication as equals. We all know that our human affections are easy to hack, so it would be easy for a superhuman AGI with almost any actual goals to trick us into liking it and make us feel that it shared our values, as exemplified in the movie *Ex Machina*.

Could any guarantees about the future behavior of the AIs, after humans are gone, make you feel good about the descendants scenario? It's a bit like writing a will for what future generations should do with our collective endowment, except that there won't be any humans around to enforce it. We'll return to the challenges of controlling the behavior of future AIs in chapter 7.

Zookeeper

Even if we get followed by the most wonderful descendants you can imagine, doesn't it feel a bit sad that there can be *no* humans left? If you prefer keeping at least some humans around no matter what, then the zookeeper scenario provides an improvement. Here an omnipotent superintelligent AI keeps some humans around, who feel treated like zoo animals and occasionally lament their fate.

Why would the zookeeper AI keep humans around? The cost of the zoo to the AI will be minimal in the grand scheme of things, and it may want to retain at least a minimal breeding population for much the same reason that we keep endangered pandas in zoos and vintage computers in museums: as an entertaining curiosity. Note that today's zoos are designed to maximize human rather than panda happiness, so we should expect human life in the zookeeper-AI scenario to be less fulfilling than it could be.

We've now considered scenarios where a free superintelligence focused on three different levels of Maslow's pyramid of human needs. Whereas the protector god AI prioritizes meaning and purpose and the benevolent dictator aims for education and fun, the zookeeper limits its attention to the lowest levels: physiological needs, safety and enough habitat enrichment to make the humans interesting to observe.

An alternate route to the zookeeper scenario is that, back when the friendly AI was created, it was designed to keep at least a billion humans safe and happy as it recursively self-improved. It has done this by confining humans to a large zoo-like happiness factory where they're kept nourished, healthy and entertained with a mixture of virtual reality and recreational drugs. The rest of Earth and our cosmic endowment are used for other purposes.

1984

If you're not 100% enthusiastic about any of the above scenarios, then consider this: Aren't things pretty nice the way they are right now, technology-wise? Can't we just keep it this way and stop worrying about AI driving us extinct or dominating us? In this spirit, let's explore a scenario where technological progress toward superintelligence is permanently curtailed not by a gatekeeper AI but by a global human-led Orwellian surveillance state where certain kinds of AI research are banned.

Technological Relinquishment

The idea of halting or relinquishing technological progress has a long and checkered history. The Luddite movement in Great Britain famously (and unsuccessfully) resisted the technology of the Industrial Revolution, and today "Luddite" is usually used as a derogatory epithet implying that someone is a technophobe on the wrong side of history, resisting progress and inevitable change. The idea of relinquishing some technologies is far from dead, however, and has found new support in the environmental and anti-globalization movements. One of its leading proponents is environmentalist Bill McKibben, who was among the first to warn of global warming. Whereas some anti-Luddites argue that all technologies should be developed and deployed so long as they're profitable, others argue that this position is too extreme, and that new technologies should be allowed only if we're confident that they'll do more good than harm. The latter is also the position of many so-called neo-Luddites.

Totalitarianism 2.0

I think that the only viable path to broad relinquishment of technology is to enforce it through a global totalitarian state. Ray Kurzweil comes to the same conclusion in *The Singularity Is Near*, as does K. Eric Drexler in *Engines of Creation*. The reason is simple economics: if some but not all relinquish a transformative technology, then the nations or groups that defect will gradually gain enough wealth and power to take over. A classic example is the British defeat of China in the First Opium War of 1839: although the Chinese invented gunpowder, they hadn't developed firearm technology as aggressively as the Europeans, and stood no chance.

Whereas past totalitarian states generally proved unstable and collapsed, novel surveillance technology offers unprecedented hope to would-be autocrats. "You know, for us, this would have been a dream come true," Wolfgang Schmidt said in a recent interview about the NSA surveillance systems revealed by Edward Snowden, recalling the days when he was a lieutenant colonel in the Stasi, the infamous secret police of East Germany.[5] Although the Stasi was often credited with building the most Orwellian surveillance state in human history, Schmidt lamented having the technology to spy on only forty phones at a time, so that adding a new citizen to the list forced him to drop another. In contrast, technology now exists that would allow a future global totalitarian state to record every phone call, email, web search, webpage view and credit card transaction for every person on Earth, and to monitor everyone's whereabouts through cell-phone tracking and surveillance cameras with face recognition. Moreover, machine learning technology far short of human-level AGI can efficiently analyze and synthesize these masses of data to identify suspected seditious behavior, enabling potential troublemakers to be neutralized before they have a chance to pose any serious challenge to the state.

Although political opposition has thus far prevented the full-scale implementation of such a system, we humans are well on our way to building the required infrastructure for the ultimate dictatorship—so in the future, when sufficiently powerful forces decided to enact this global 1984 scenario, they found that they didn't need to do much more than flip the on switch. Just as in George Orwell's novel *Nineteen Eighty-Four*, the ultimate power in this future global state resides not with a traditional dictator, but with the human-made bureaucratic system itself. There is no single person who is extraordinarily powerful; rather, all are pawns in a chess game whose draconian rules nobody is able to change or challenge. By engineering a system where people keep one another in check with the surveillance technology, this faceless, leaderless state is able to last for many millennia, keeping Earth free from superintelligence.

Discontent

This society, of course, lacks all the benefits that only superintelligence-enabled technology can bring. Most people don't lament this because they don't know what they're missing: the whole idea of superintelligence has long since been deleted from the official historical records, and advanced AI research is banned. Every so often, a freethinker is born who dreams of a more open and dynamic society where knowledge can grow and rules can be changed. However, the only ones who last long are the ones who learn to keep these ideas strictly to themselves, flickering alone like transient sparks without ever starting a fire.

Reversion

Wouldn't it be tempting to escape the perils of technology without succumbing to stagnant totalitarianism? Let's explore a scenario where this was accomplished by reverting to primitive technology, inspired by the Amish. After the Omegas took over the world as in the opening of the book, a massive global propaganda campaign was launched that romanticized the simple farming life of 1,500 years ago. Earth's population was reduced to about 100 million people by an engineered pandemic blamed on terrorists. The pandemic was secretly targeted to ensure that nobody who knew anything about science or technology survived. With the excuse of eliminating the infection hazard of large concentrations of people, Prometheus-controlled robots emptied and razed all cities. Survivors were given large tracts of (suddenly available) land and educated in sustainable farming, fishing and hunting practices using only early medieval technology. In the meantime, armies of robots systematically removed all traces of modern technology (including cities, factories, power lines and paved roads), and thwarted all human attempts to document or re-create any such technology. Once the technology was globally forgotten, robots helped dismantle other robots until there were almost none left. The very last robots were deliberately vaporized together with Prometheus itself in a large thermonuclear explosion. There was no longer any need to ban modern technology, since it was all gone. As a result, humanity bought itself over a millennium of additional time without worries about either AI or totalitarianism.

Reversion has to a lesser extent happened before: for example, some of the technologies that were in widespread use during the Roman Empire were largely forgotten for about a millennium before making a comeback during the Renaissance. Isaac Asimov's *Foundation* trilogy centers around the "Seldon Plan" to shorten a reversion period from 30,000 years to 1,000 years. With clever planning, it may be possible to do the opposite and lengthen rather than shorten a reversion period, for example by erasing all knowledge of agriculture. However, unfortunately for reversion enthusiasts, it's unlikely that this scenario can be extended indefinitely without humanity either going high-tech or going extinct. Counting on people's resembling today's biological humans 100 million years from now would be naive, given that we haven't existed as a species for more than 1% of that time so far. Moreover, low-tech humanity would be a defenseless sitting duck just waiting to be exterminated by the next planet-scorching asteroid impact or other mega-calamity brought on by Mother Nature. We certainly can't last a billion years, after which the gradually warming

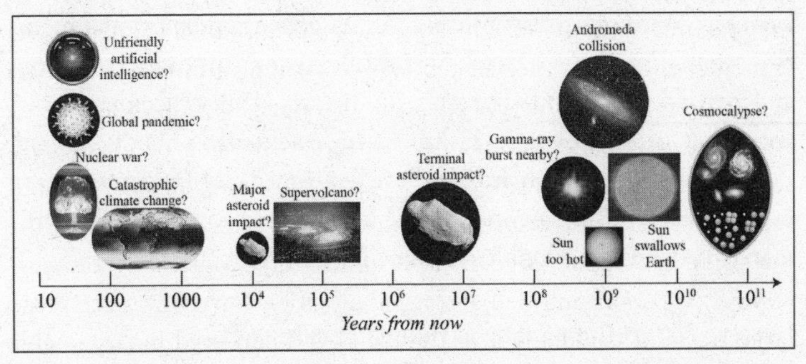

Figure 5.1: Examples of what could destroy life as we know it or permanently curtail its potential. Whereas our Universe itself will likely last for at least tens of billions of years, our Sun will scorch Earth in about a billion years and then swallow it unless we move it a safe distance, and our Galaxy will collide with its neighbor in about 3.5 billion years. Although we don't know exactly when, we can predict with near certainty that long before this, asteroids will pummel us and supervolcanoes will cause year-long sunless winters. We can use technology either to solve all these problems or to create new ones such as climate change, nuclear war, engineered pandemics or AI gone awry.

Sun will have cranked up Earth's temperature enough to boil off all liquid water.

Self-Destruction

After contemplating problems that future technology might cause, it's important to also consider problems that *lack of* that technology can cause. In this spirit, let us explore scenarios where superintelligence is never created because humanity eliminates itself by other means.

How might we accomplish that? The simplest strategy is "just wait." Although we'll see in the next chapter how we can solve such problems as asteroid impacts and boiling oceans, these solutions all require technology that we haven't yet developed, so unless our technology advances far beyond its present level, Mother Nature will drive us extinct long before another billion years have passed. As the famous economist John Maynard Keynes said: "In the long run we are all dead."

Unfortunately, there are also ways in which we might self-destruct much sooner, through collective stupidity. Why would our species commit collective suicide, also known as *omnicide*, if virtually nobody wants it? With our present level of intelligence and emotional maturity, we humans have a knack for miscalculations, misunderstandings and incompetence, and as a result, our history is full of accidents, wars and other calamities that, in hindsight, essentially nobody wanted. Economists and mathematicians have developed elegant game-theory explanations for how people can be incentivized to actions that ultimately cause a catastrophic outcome for everyone.[6]

Nuclear War: A Case Study in Human Recklessness

You might think that the greater the stakes, the more careful we'd be, but a closer examination of the greatest risk that our current technology permits, namely a global thermonuclear war, isn't reassuring. We've had to rely on luck to weather an embarrassingly long list of near misses caused by all sorts of things: computer malfunction, power failure, faulty intelligence, navigation error, bomber crash, satellite explosion and so on.[7] In fact, if it weren't for heroic acts of certain individuals—for example, Vasili Arkhipov and Stanislav Petrov—we might already have had a global nuclear war. Given our track record, I think it's highly unlikely that the annual probability of accidental nuclear war is as low as one in a thousand if we keep up our present behavior, in which case the probability that we'll have one within 10,000 years exceeds $1 - 0.999^{10000} \approx 99.995\%$.

To fully appreciate our human recklessness, we must realize that we started the nuclear gamble even before carefully studying the risks. First, radiation risks had been underestimated, and over $2 billion in compensation has been paid out to victims of radiation exposure from uranium handling and nuclear tests in the United States alone.[8]

Second, it was eventually discovered that hydrogen bombs deliberately detonated hundreds of kilometers above Earth would create a powerful electromagnetic pulse (EMP) that might disable the electric grid and electronic devices over vast areas (figure 5.2), leaving infrastructure paralyzed, roads clogged with disabled vehicles and conditions for nuclear-aftermath survival less than ideal. For example, the U.S. EMP Commission reported that "the water infrastructure is a vast machine, powered partly by gravity but mostly by electricity," and that denial of water can cause death in three to four days.[9]

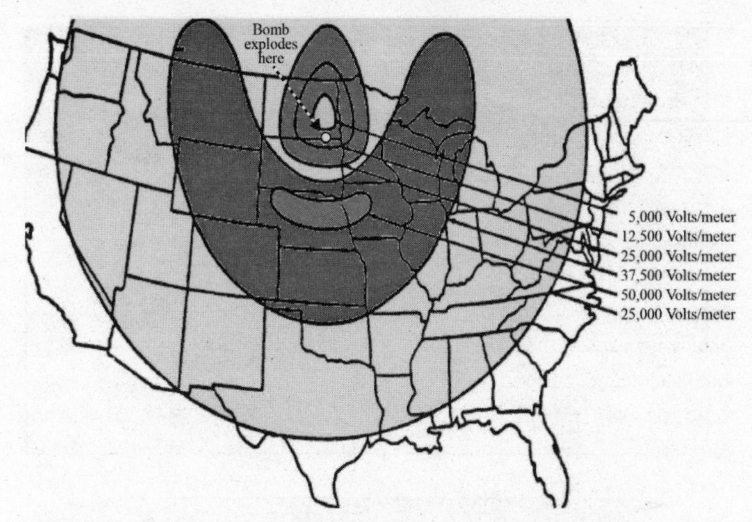

Figure 5.2: A single hydrogen bomb explosion 400 km above Earth can cause a powerful electromagnetic pulse that can cripple electricity-using technology over a vast area. By shifting the detonation point southeast, the banana-shaped zone exceeding 37,500 volts per meter could cover most of the U.S. East Coast. Reprinted from U.S. Army Report AD-A278230 (unclassified) with colors added.

Third, the potential of nuclear winter wasn't realized until four decades in, after we'd deployed 63,000 hydrogen bombs—oops! Regardless of whose cities burned, massive amounts of smoke reaching the upper troposphere might spread around the globe, blocking out enough sunlight to transform summers into winters, much like when an asteroid or supervolcano caused a mass extinction in the past. When the alarm was sounded by both U.S. and Soviet scientists in the 1980s, this contributed to the decision of Ronald Reagan and Mikhail Gorbachev to start slashing stockpiles.[10] Unfortunately, more accurate calculations have painted an even gloomier picture: figure 5.3 shows cooling by about 20° Celsius (36° Fahrenheit) in much of the core farming regions of the United States, Europe, Russia and China (and by 35°C in some parts of Russia) for the first two summers, and about half that even a full decade later.* What does that mean in plain

* Injecting carbon into the atmosphere can cause two kinds of climate change: warming from carbon dioxide or cooling from smoke and soot. It's not only the first kind that's occasionally dismissed without scientific evidence: I'm sometimes told that

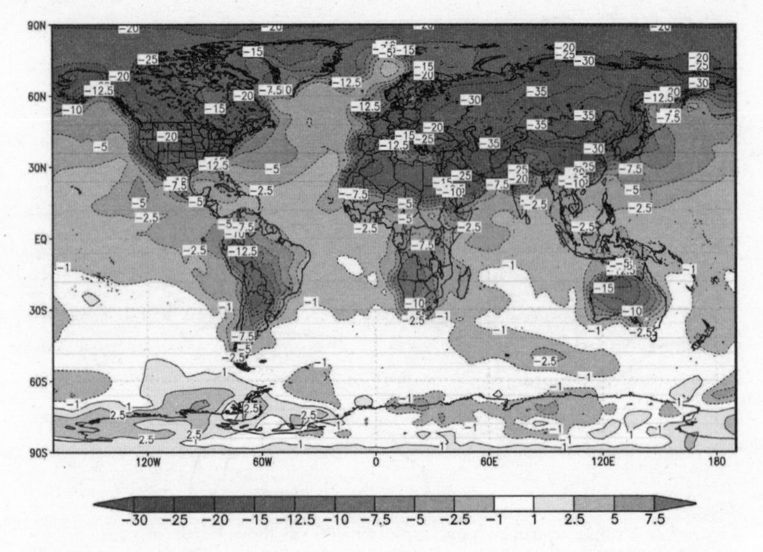

Figure 5.3: Average cooling (in °C) during the first two summers after a full-scale nuclear war between the United States and Russia. Reproduced with permission from Alan Robock.[11]

English? One doesn't need much farming experience to conclude that near-freezing summer temperatures for years would eliminate most of our food production. It's hard to predict exactly what would happen after thousands of Earth's largest cities are reduced to rubble and global infrastructure collapses, but whatever small fraction of all humans don't succumb to starvation, hypothermia or disease would need to cope with roving armed gangs desperate for food.

I've gone into such detail on global nuclear war to drive home the crucial point that no reasonable world leader would want it, yet it might nonetheless happen by accident. This means that we can't trust our fellow humans never to commit omnicide: nobody wanting it isn't necessarily enough to prevent it.

---

nuclear winter has been debunked and is virtually impossible. I always respond by asking for a reference to a peer-reviewed scientific paper making such strong claims and, so far, there seem to be none whatsoever. Although there are great uncertainties that warrant further research, especially related to how much smoke gets produced and how high up it rises, there's in my scientific opinion no current basis for dismissing the nuclear winter risk.

Doomsday Devices

So could we humans actually pull off omnicide? Even if a global nuclear war may kill off 90% of all humans, most scientists guess that it wouldn't kill 100% and therefore wouldn't drive us extinct. On the other hand, the story of nuclear radiation, nuclear EMP and nuclear winter all demonstrate that the greatest hazards may be ones we haven't even thought of yet. It's incredibly difficult to foresee all aspects of the aftermath, and how nuclear winter, infrastructure collapse, elevated mutation levels and desperate armed hordes might interact with other problems such as new pandemics, ecosystem collapse and effects we haven't yet imagined. My personal assessment is therefore that although the probability of a nuclear war tomorrow triggering human extinction isn't large, we can't confidently conclude that it's zero either.

Omnicide odds increase if we upgrade today's nuclear weapons into a deliberate doomsday device. Introduced by RAND strategist Herman Kahn in 1960 and popularized in Stanley Kubrick's film *Dr. Strangelove*, a doomsday device takes the paradigm of mutually assured destruction to its ultimate conclusion. It's the perfect deterrent: a machine that automatically retaliates against any enemy attack by killing all of humanity.

One candidate for the doomsday device is a huge underground cache of so-called *salted nukes*, preferably humongous hydrogen bombs surrounded by massive amounts of cobalt. Physicist Leo Szilard argued already in 1950 that this could kill everyone on Earth: the hydrogen bomb explosions would render the cobalt radioactive and blow it into the stratosphere, and its five-year half-life is long enough for it to settle all across Earth (especially if twin doomsday devices were placed in opposite hemispheres), but short enough to cause lethal radiation intensity. Media reports suggest that cobalt bombs are now being built for the first time. Omnicidal opportunities could be bolstered by adding bombs optimized for nuclear winter creation by maximizing long-lived aerosols in the stratosphere. A major selling point of a doomsday device is that it's much cheaper than a conventional nuclear deterrent: since the bombs don't need to be launched, there's no need for expensive missile systems, and the bombs themselves are cheaper to build since they need not be light and compact enough to fit into missiles.

Another possibility is the future discovery of a biological doomsday device: a custom-designed bacterium or virus that kills all humans. If its transmissibility were high enough and its incubation period long enough, essentially everybody could catch it before they realized its existence and took countermeasures. There's a military argument for building such a bioweapon even if it can't kill everybody: the most effective doomsday device is one that combines nuclear, biological and other weapons to maximize the chances of deterring the enemy.

AI Weapons

A third technological route to omnicide may involve relatively dumb AI weapons. Suppose a superpower builds billions of those bumblebee-sized attack drones from chapter 3 and uses them to kill anyone except their own citizens and allies, identified remotely by a radio-frequency ID tag just as most of today's supermarket products. These tags could be distributed to all citizens to be worn on bracelets or as transdermal implants, as in the totalitarianism section. This would probably spur an opposing superpower to build something analogous. When war accidentally breaks out, all humans would be killed, even unaffiliated remote tribes, because nobody would be wearing both kinds of ID tag. Combining this with a nuclear and biological doomsday device would further improve chances of successful omnicide.

What Do *You* Want?

You began this chapter pondering where you want the current AGI race to lead. Now that we've explored a broad range of scenarios together, which ones appeal to you and which ones do you think we should try hard to avoid? Do you have a clear favorite? Please let me and fellow readers know at http://AgeOfAi.org, and join the discussion!

The scenarios we've covered obviously shouldn't be viewed as a complete list, and many are thin on details, but I've tried hard to be inclusive, spanning the full spectrum from high-tech to low-tech to no-tech and describing all the central hopes and fears expressed in the literature.

One of the most fun parts of writing this book has been hearing what my friends and colleagues think of these scenarios, and I've been amused to learn that there's no consensus whatsoever. The one thing everybody agrees on is that the choices are more subtle than they may initially seem. People who like any one scenario tend to simultaneously find some aspect(s) of it bothersome. To me, this means that we humans need to continue and deepen this conversation about our future goals, so that we know in which direction to steer. The future potential for life in our cosmos is awe-inspiringly grand, so let's not squander it by drifting like a rudderless ship, clueless about where we want to go!

Just how grand is this future potential? No matter how advanced our technology gets, the ability for Life 3.0 to improve and spread through our cosmos will be limited by the laws of physics—what are these ultimate limits, during the billions of years to come? Is our Universe teeming with extraterrestrial life right now, or are we alone? What happens if different expanding cosmic civilizations meet? We'll tackle these fascinating questions in the next chapter.

THE BOTTOM LINE:

- The current race toward AGI can end in a fascinatingly broad range of aftermath scenarios for upcoming millennia.
- Superintelligence can peacefully coexist with humans either because it's forced to (enslaved-god scenario) or because it's "friendly AI" that wants to (libertarian-utopia, protector-god, benevolent-dictator and zookeeper scenarios).
- Superintelligence can be prevented by an AI (gatekeeper scenario) or by humans (1984 scenario), by deliberately forgetting the technology (reversion scenario) or by lack of incentives to build it (egalitarian-utopia scenario).
- Humanity can go extinct and get replaced by AIs (conqueror and descendant scenarios) or by nothing (self-destruction scenario).
- There's absolutely no consensus on which, if any, of these scenarios are desirable, and all involve objectionable elements. This makes it all the more important to continue and deepen the conversation around our future goals, so that we don't inadvertently drift or steer in an unfortunate direction.

Chapter 6

Our Cosmic Endowment: The Next Billion Years and Beyond

Our speculation ends in a supercivilization, the synthesis of all solar-system life, constantly improving and extending itself, spreading outward from the sun, converting nonlife into mind.

Hans Moravec, *Mind Children*

To me, the most inspiring scientific discovery ever is that we've dramatically underestimated life's future potential. Our dreams and aspirations need not be limited to century-long life spans marred by disease, poverty and confusion. Rather, aided by technology, life has the potential to flourish for billions of years, not merely here in our Solar System, but also throughout a cosmos far more grand and inspiring than our ancestors imagined. Not even the sky is the limit.

This is exciting news for a species that has been inspired by pushing limits throughout the ages. Olympic games celebrate pushing the limits of strength, speed, agility and endurance. Science celebrates pushing the limits of knowledge and understanding. Literature and art celebrate pushing the limits of creating beautiful or life-enriching experiences. Many people, organizations and nations celebrate increasing resources, territory and longevity. Given our human obsession with limits, it's fitting that the best-selling copyrighted book of all time is *The Guinness Book of World Records*.

So if our old perceived limits of life can be shattered by technology, what are the *ultimate* limits? How much of our cosmos can come alive? How far can life reach and how long can it last? How much matter can life make use of, and how much energy, information and computation can it extract? These ultimate limits are set not by our understanding, but by the laws of physics. This, ironically, makes it in some ways easier to analyze the long-term future of life than the short-term future.

If our 13.8-billion-year cosmic history were compressed into a week, then the 10,000-year drama of the last two chapters would be over in less than half a second. This means that although we cannot predict if and how an intelligence explosion will unfold and what its immediate aftermath will be like, all this turmoil is merely a brief flash in cosmic history whose details don't affect life's ultimate limits. If the post-explosion life is as obsessed as today's humans are with pushing limits, then it will develop technology to actually *reach* these limits—because it can. In this chapter, we'll explore what these limits are, thus getting a glimpse of what the long-term future of life may be like. Since these limits are based on our current understanding of physics, they should be viewed as a lower bound on the possibilities: future scientific discoveries may present opportunities to do even better.

But do we really know that future life will be so ambitious? No, we don't: perhaps it will become as complacent as a heroin addict or a couch potato merely watching endless reruns of *Keeping Up with the Kardashians*. However, there is reason to suspect that ambition is a rather generic trait of advanced life. Almost regardless of what it's trying to maximize, be it intelligence, longevity, knowledge or interesting experiences, it will need resources. It therefore has an incentive to push its technology to the ultimate limits, to make the most of the resources it has. After this, the only way to further improve is to acquire more resources, by expanding into ever-larger regions of the cosmos.

Also, life may independently originate in multiple places in our cosmos. In that case, unambitious civilizations simply become cosmically irrelevant, with ever-larger parts of the cosmic endowment ultimately being taken over by the most ambitious life forms. Natural selection therefore plays out on a cosmic scale and, after a while, almost all life that exists will be ambitious life. In summary, if we're interested in the extent to which our cosmos can ultimately come alive, we should study the limits of ambition that are imposed by the laws of physics. Let's do this! Let's first explore the limits of what can be done with the resources (matter, energy, etc.) that we have in our Solar System, then turn to how to get more resources through cosmic exploration and settlement.

Making the Most of Your Resources

Whereas today's supermarkets and commodity exchanges sell tens of thousands of items we might call "resources," future life that's reached the technological limit needs mainly one fundamental resource: so-called *baryonic matter*, meaning anything made up of atoms or their constituents (quarks and electrons). Whatever form this matter is in, advanced technology can rearrange it into any desired substances or objects, including power plants, computers and advanced life forms. Let's therefore begin by examining the limits on the energy that powers advanced life and the information processing that enables it to think.

Building Dyson Spheres

When it comes to the future of life, one of the most hopeful visionaries is Freeman Dyson. I've had the honor and pleasure of knowing him for the past two decades, but when I first met him, I felt nervous. I was a junior postdoc chowing away with my friends in the lunchroom of the Institute for Advanced Study in Princeton, and out of the blue, this world-famous physicist who used to hang out with Einstein and Gödel came up and introduced himself, asking if he could join us! He quickly put me at ease, however, by explaining that he preferred eating lunch with young folks over stuffy old professors. Even though he's ninety-three as I type these words, Freeman is still younger in spirit than most people I know, and the mischievous boyish glint in his eyes reveals that he couldn't care less about formalities, academic hierarchies or conventional wisdom. The bolder the idea, the more excited he gets.

When we talked about energy use, he scoffed at how unambitious we humans were, pointing out that we could meet all our current global energy needs by harvesting the sunlight striking an area smaller than 0.5% of the Sahara desert. But why stop there? Why even stop at capturing all the sunlight striking Earth, letting most of it get wastefully beamed into empty space? Why not simply put *all* the Sun's energy output to use for life?

Inspired by Olaf Stapledon's 1937 sci-fi classic *Star Maker*, with rings of artificial worlds orbiting their parent star, Freeman Dyson published a description in 1960 of what became known as a *Dyson sphere*.[1] Freeman's idea was to rearrange Jupiter into a biosphere in the form of a spherical shell surrounding the Sun, where our descendants could flourish, enjoying 100 billion times more biomass and a trillion times more energy than humanity uses today.[2] He argued that this was the natural next step: "One should expect that, within a few thousand years of its entering the stage of industrial development, any intelligent species should be found occupying an artificial biosphere which completely surrounds its parent star." If you lived on the inside of a Dyson sphere, there would be no nights: you'd always see the Sun straight overhead, and all across the sky, you'd see sunlight reflecting off the rest of the biosphere, just as you can nowadays see sunlight reflecting off the Moon during the day. If you wanted to see stars, you'd simply go "upstairs" and peer out at the cosmos from the outside of the Dyson sphere.

A low-tech way to build a partial Dyson sphere is to place a ring of habitats in circular orbit around the Sun. To completely surround the Sun, you could add rings orbiting it around different axes at slightly different distances, to avoid collisions. To avoid the nuisance that these fast-moving rings couldn't be connected to one another, complicating transportation and communication, one could instead build a monolithic stationary Dyson sphere where the Sun's inward gravitational pull is balanced by the outward pressure from the Sun's radiation—an idea pioneered by Robert L. Forward and by Colin McInnes. The sphere can be built by gradually adding more "statites": stationary satellites that counteract the Sun's gravity with radiation pressure rather than centrifugal forces. Both of these forces drop off with the square of the distance to the Sun, which means that if they can be balanced at one distance from the Sun, they'll conveniently be balanced at any other distance as well, allowing freedom to park anywhere in our Solar System. Statites need to be extremely lightweight sheets, weighing only 0.77 grams per square meter, which is about

100 times less than paper, but this is unlikely to be a showstopper. For example, a sheet of graphene (a single layer of carbon atoms in a hexagonal pattern resembling chicken wire) weighs a thousand times less than that limit. If the Dyson sphere is built to reflect rather than absorb most of the sunlight, then the total intensity of light bouncing around within it will be dramatically increased, further boosting the radiation pressure and the amount of mass that can be supported in the sphere. Many other stars have a thousandfold and even a millionfold greater luminosity than our Sun, and are therefore able to support correspondingly heavier stationary Dyson spheres.

If a much heavier rigid Dyson sphere is desired here in our Solar System, then resisting the Sun's gravity will require ultra-strong materials that can withstand pressures tens of thousands of times greater than those at the base of the world's tallest skyscrapers, without liquefying or buckling. To be long-lived, a Dyson sphere would need to be dynamic and intelligent, constantly fine-tuning its position and shape in response to disturbances and occasionally opening up large holes to let annoying asteroids and comets pass through without incident. Alternatively, a detect-and-deflect system could be used to handle such system intruders, optionally disassembling them and putting their matter to better use.

For today's humans, life on or in a Dyson sphere would at best be disorienting and at worst impossible, but that need not stop future biological or non-biological life forms from thriving there. The orbiting variant would offer essentially no gravity at all, and if you walked around on the stationary kind, you could walk only on the outside (facing away from the Sun) without falling off, with gravity about ten thousand times weaker than you're used to. You'd have no magnetic field (unless you built one) shielding you from dangerous particles from the Sun. The silver lining is that a Dyson sphere the size of Earth's current orbit would give us about 500 million times more surface area to live on.

If more Earth-like human habitats are desired, the good news is that they're much easier to build than a Dyson sphere. For example, figures 6.1 and 6.2 show a cylindrical habitat design pioneered by the American physicist Gerard K. O'Neill, which supports artificial gravity, cosmic ray shielding, a twenty-four-hour day-night cycle,

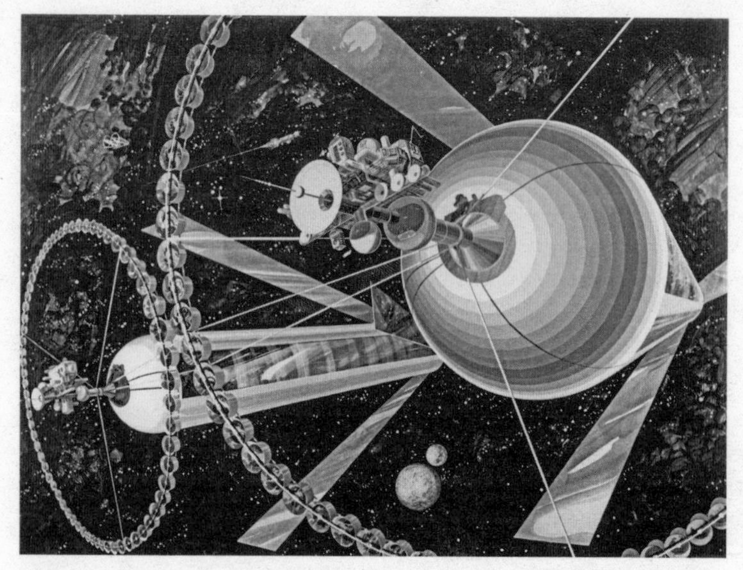

Figure 6.1: A pair of counterrotating O'Neill cylinders can provide comfortable Earth-like human habitats if they orbit the Sun in such a way that they always point straight at it. The centrifugal force from their rotation provides artificial gravity, and three foldable mirrors beam sunlight inside on a 24-hour day-night cycle. The smaller habitats arranged in a ring are specialized for agriculture. Image courtesy of Rick Guidice/NASA.

and Earth-like atmosphere and ecosystems. Such habitats could orbit freely inside a Dyson sphere, or modified variants could be attached outside it.

Building Better Power Plants

Although Dyson spheres are energy efficient by today's engineering standards, they come nowhere near pushing the limits set by the laws of physics. Einstein taught us that if we could convert mass to energy with 100% efficiency,* then an amount of mass m would give us an amount of energy E given by his famous formula $E = mc^2$, where c is the speed of light. This means that since c is huge, a small amount of

* If you work in the energy sector, you may be used to instead defining efficiency as the fraction of the energy released that's in a useful form.

Figure 6.2: Interior view of one of the O'Neill cylinders from the previous figure. If its diameter is 6.4 kilometers and rotates once every 2 minutes, people on the surface will experience the same apparent gravity as on Earth. The Sun is behind you, but appears above because of a mirror outside the cylinder that folds away at night. Airtight windows keep the atmosphere from escaping the cylinder. Image courtesy of Rick Guidice/NASA.

mass can produce a humongous amount of energy. If we had an abundant supply of antimatter (which we don't), then a 100% efficient power plant would be easy to make: simply pouring a teaspoonful of anti-water into regular water would unleash the energy equivalent to 200,000 tons of TNT, the yield of a typical hydrogen bomb—enough to power the world's entire energy needs for about seven minutes.

In contrast, our most common ways of generating energy today are woefully inefficient, as summarized in table 6.1 and figure 6.3. Digesting a candy bar is merely 0.00000001% efficient, in the sense that it releases a mere ten-trillionth of the energy mc^2 that it contains. If your stomach were even 0.001% efficient, then you'd only need to eat a single meal for the rest of your life. Compared to eating, the burning of coal and gasoline are merely 3 and 5 times more efficient,

Method	Efficiency
Digesting candy bar	0.00000001%
Burning coal	0.00000003%
Burning gasoline	0.00000005%
Fission of uranium-235	0.08%
Using Dyson sphere until Sun dies	0.08%
Fusion of hydrogen to helium	0.7%
Spinning black hole engine	29%
Dyson sphere around quasar	42%
Sphalerizer	50%?
Black hole evaporation	90%

Table 6.1: Efficiency of converting mass into usable energy relative to the theoretical limit $E = mc^2$. As explained in the text, getting 90% efficiency from feeding black holes and waiting for them to evaporate is unfortunately too slow to be useful, and accelerating the process dramatically lowers the efficiency.

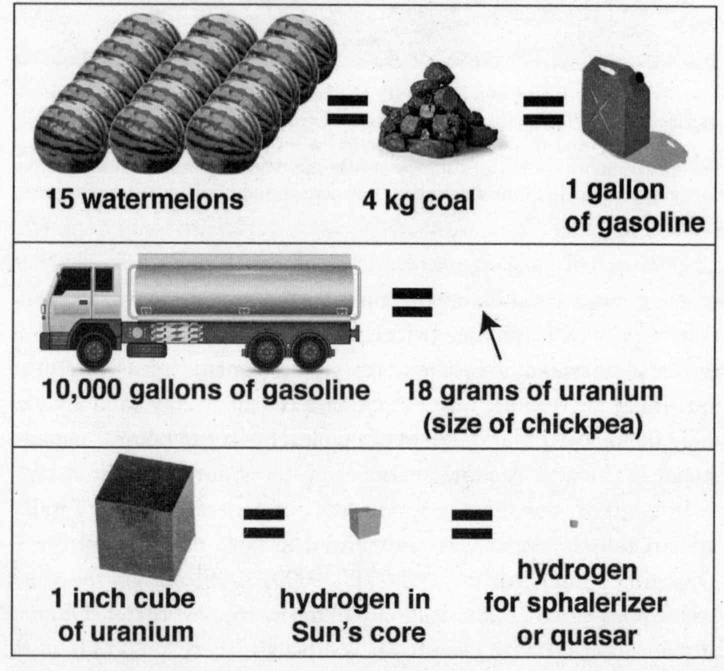

Figure 6.3: Advanced technology can extract dramatically more energy from matter than we get by eating or burning it, and even nuclear fusion extracts 140 times less energy than the limits set by the laws of physics. Power plants exploiting sphalerons, quasars or evaporating black holes might do much better.

respectively. Today's nuclear reactors do dramatically better by splitting uranium atoms through fission, but still fail to extract more than 0.08% of their energy. The nuclear reactor in the core of the Sun is an order of magnitude more efficient than those we've built, extracting 0.7% of the energy from hydrogen by fusing it into helium. However, even if we enclose the Sun in a perfect Dyson sphere, we'll never convert more than about 0.08% of the Sun's mass to energy we can use, because once the Sun has consumed about about a tenth of its hydrogen fuel, it will end its lifetime as a normal star, expand into a red giant, and begin to die. Things don't get much better for other stars either: the fraction of their hydrogen consumed during the main lifetime ranges from about 4% for very small stars to about 12% for the largest ones. If we perfect an artificial fusion reactor that would let us fuse 100% of all hydrogen at our disposal, we'd still be stuck at that embarrassingly low 0.7% efficiency of the fusion process. How can we do better?

Evaporating Black Holes

In his book *A Brief History of Time*, Stephen Hawking proposed a black hole power plant.* This may sound paradoxical given that black holes were long believed to be traps that nothing, not even light, could ever escape from. However, Hawking famously calculated that quantum gravity effects make a black hole act like a hot object—the smaller, the hotter—that gives off heat radiation now known as *Hawking radiation*. This means that the black hole gradually loses energy and evaporates away. In other words, whatever matter you dump into the black hole will eventually come back out again as heat radiation, so by the time the black hole has completely evaporated, you've converted your matter to radiation with nearly 100% efficiency.†

* If no suitable nature-made black hole can be found in the nearby universe, a new one can be created by putting lots of matter in a sufficiently small space.

† This is a slight oversimplification, because Hawking radiation also includes some particles from which it's hard to extract useful work. Large black holes are only 90% efficient, because about 10% of the energy is radiated in the form of gravitons: extremely shy particles that are almost impossible to detect, let alone extract useful work from. As the black hole continues evaporating and shrinking, the efficiency drops further because the Hawking radiation starts including neutrinos and other massive particles.

A problem with using black hole evaporation as a power source is that, unless the black hole is much smaller than an atom in size, it's an excruciatingly slow process that takes longer than the present age of our Universe and radiates less energy than a candle. The power produced decreases with the square of the size of the hole, and the physicists Louis Crane and Shawn Westmoreland have therefore proposed using a black hole about a thousand times smaller than a proton, weighing about as much as the largest-ever seagoing ship.[3] Their main motivation was to use the black hole engine to power a starship (a topic to which we return below), so they were more concerned with portability than efficiency and proposed feeding the black hole with laser light, causing no energy-to-matter conversion at all. Even if you could feed it with matter instead of radiation, guaranteeing high efficiency appears difficult: to make protons enter such a black hole a thousandth their size, they would have to be fired at the hole with a machine as powerful as the Large Hadron Collider, augmenting their energy mc^2 with at least a thousand times more kinetic (motion) energy. Since at least 10% of that kinetic energy would be lost to gravitons when the black hole evaporates, we'd therefore be putting more energy into the black hole than we'd be able to extract and put to work, ending up with negative efficiency. Further confounding the prospects of a black hole power plant is that we still lack a rigorous theory of quantum gravity upon which to base our calculations—but this uncertainty could of course also mean that there are new useful quantum gravity effects yet to be discovered.

Spinning Black Holes

Fortunately, there are other ways of using black holes as power plants that don't involve quantum gravity or other poorly understood physics. For example, many existing black holes spin very fast, with their event horizons whirling around near the speed of light, and this rotation energy can be extracted. The event horizon of a black hole is the region from which not even light can escape, because the gravitational pull is too powerful. Figure 6.4 illustrates how outside the event horizon, a spinning black hole has a region called the *ergosphere*, where the spinning black hole drags space along with it so fast that it's impossible for a particle to sit still and not get dragged along. If you toss an

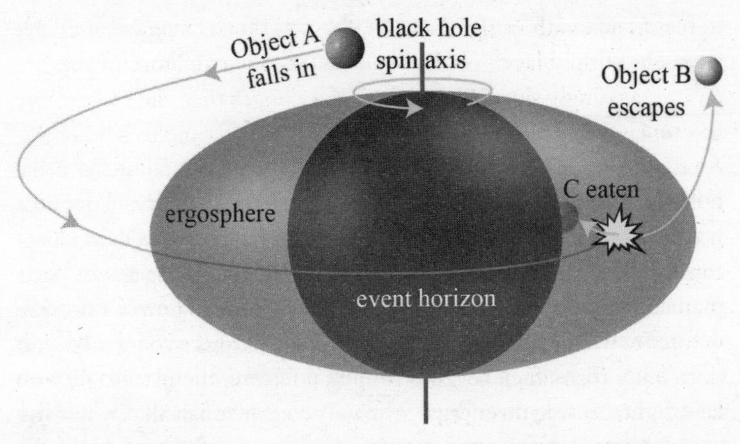

Figure 6.4: Part of the rotational energy of a spinning black hole can be extracted by throwing a particle A near the black hole and having it split into a part C that gets eaten and a part B that escapes—with more energy than A had initially.

object into the ergosphere, it will therefore pick up speed rotating around the hole. Unfortunately, it will soon get eaten up by the black hole, forever disappearing through the event horizon, so this does you no good if you're trying to extract energy. However, Roger Penrose discovered that if you launch the object at a clever angle and make it split into two pieces as figure 6.4 illustrates, then you can arrange for only one piece to get eaten while the other escapes the black hole with more energy than you started with. In other words, you've successfully converted some of the rotational energy of the black hole into useful energy that you can put to work. By repeating this process many times, you can milk the black hole of *all* its rotational energy so that it stops spinning and its ergosphere disappears. If the initial black hole was spinning as fast as nature allows, with its event horizon moving essentially at the speed of light, this strategy allows you to convert 29% of its mass into energy. There is still significant uncertainty about how fast the black holes in our night sky spin, but many of the best-studied ones appear to spin quite fast: between 30% and 100% of the maximum allowed. The monster black hole in the middle of our Galaxy (which weighs four million times as much as our Sun) appears to spin, so even if only 10% of its mass could be converted to useful energy, that would deliver the same as 400,000 suns converted to energy with 100% efficiency, or about as much energy as we'd get from Dyson spheres around 500 million suns over billions of years.

Quasars

Another interesting strategy is to extract energy not from the black hole itself, but from matter falling into it. Nature has already found a way of doing this all on its own: the quasar. As gas swirls even closer to a black hole, forming a pizza-shaped disk whose innermost parts gradually get gobbled up, it gets extremely hot and gives off copious amounts of radiation. As gas falls downward toward the hole, it speeds up, converting its gravitational potential energy into motion energy, just as a skydiver does. The motion gets progressively messier as complicated turbulence converts the coordinated motion of the gas blob into random motion on ever-smaller scales, until individual atoms begin colliding with each other at high speeds—having such random motion is precisely what it means to be hot, and these violent collisions convert motion energy into radiation. By building a Dyson sphere around the entire black hole, at a safe distance, this radiation energy can be captured and put to use. The faster the black hole spins, the more efficient this process gets, with a maximally spinning black hole delivering energy at a whopping 42% efficiency.* For black holes weighing about as much as a star, most of the energy comes out as X-rays, whereas for the supermassive kind found in the centers of galaxies, much of it emerges somewhere in the range of infrared, visible and ultraviolet light.

Once you've run out of fuel to feed your black hole, you can switch to extracting its rotational energy as we discussed above.† Indeed, nature has already found a way of partially doing that as well, boosting the radiation from accreted gas through a magnetic process

* For Douglas Adams fans out there, note that this is an elegant question giving the answer to the question of life, the universe and everything. More precisely, the efficiency is $1 - 1/\sqrt{3} \approx 42\%$.

† If you feed the black hole by placing a gas cloud around it that rotates slowly in the same direction, then this gas will spin ever faster as it's pulled in and eaten, boosting the black hole's rotation, just as a figure-skater spins faster when pulling in her arms. This may keep the hole maximally spinning, enabling you to extract first 42% of the gas energy and then 29% of the remainder, for a total efficiency of $42\% + (1-42\%)\times 29\% \approx 59\%$.

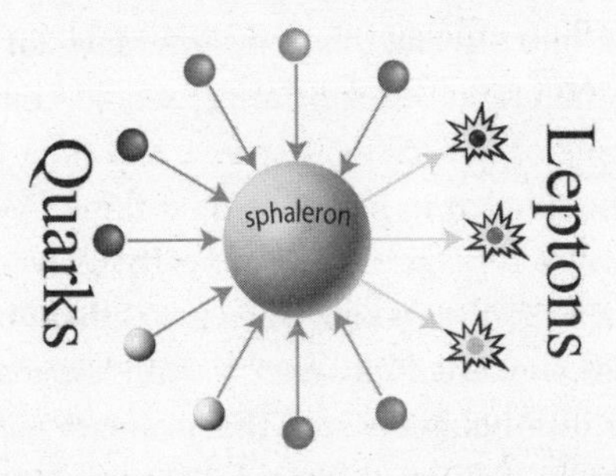

Figure 6.5: According to the standard model of particle physics, nine quarks with appropriate flavor and spin can come together and transform into three leptons through an intermediate state called a sphaleron. The combined mass of the quarks (together with the energy of the gluon particles that accompanied them) is much greater than the mass of the leptons, so this process will release energy, indicated by flashes.

known as the Blandford-Znajek mechanism. It may well be possible to use technology to further improve the energy extraction efficiency beyond 42% by clever use of magnetic fields or other ingredients.

Sphalerons

There is another known way to convert matter into energy that doesn't involve black holes at all: the *sphaleron* process. It can destroy quarks and turn them into leptons: electrons, their heavier cousins the muon and tau particles, neutrinos or their antiparticles.[4] As illustrated in figure 6.5, the standard model of particle physics predicts that nine quarks with appropriate flavor and spin can come together and transform into three leptons through an intermediate state called a sphaleron. Because the input weighs more than the output, the mass difference gets converted into energy according to Einstein's $E = mc^2$ formula.

Future intelligent life might therefore be able to build what I'll call a *sphalerizer:* an energy generator acting like a diesel engine on steroids. A traditional diesel engine compresses a mixture of air and diesel oil until the temperature gets high enough for it to spontaneously ignite and burn, after which the hot mixture re-expands and does useful work in the process, say pushing a piston. The carbon dioxide and other combustion gases weigh about 0.00000005% less than what was in the piston initially, and this mass difference turns into the heat energy driving the engine. A sphalerizer would compress ordinary matter to a couple of quadrillion degrees, and then let it re-expand and cool once the sphalerons had done their thing.* We already know the result of this experiment, because our early Universe performed it for us about 13.8 billion years ago, when it was that hot: almost 100% of the matter gets converted into energy, with less than a billionth of the particles left over being the stuff that ordinary matter is made of: quarks and electrons. So it's just like a diesel engine, except over a billion times more efficient! Another advantage is that you don't need to be finicky about what to fuel it with—it works with anything made of quarks, meaning any normal matter at all.

Because of these high-temperature processes, our baby Universe produced over a trillion times more radiation (photons and neutrinos) than matter (quarks and electrons that later clumped into atoms). During the 13.8 billion years since then, a great segregation took place, where atoms became concentrated into galaxies, stars and planets, while most photons stayed in intergalactic space, forming the cosmic microwave background radiation that has been used to make baby pictures of our Universe. Any advanced life form living in a galaxy or other matter concentration can therefore turn most of its available matter back into energy, rebooting the matter percentage down to the same tiny value that emerged from our early Universe by briefly re-creating those hot dense conditions inside a sphalerizer.

To figure out how efficient an actual sphalerizer would be, one needs to work out key practical details: for example, how large does it need to be to prevent a significant fraction of the photons and neutrinos from leaking out during the compression stage? What we can say for sure, however, is that the energy prospects for the future of life are dramatically better than our current technology allows. We haven't even managed to build a fusion reactor, yet future technology should be able to do ten and perhaps even a hundred times better.

* It needs to get hot enough to re-unify the electromagnetic and weak forces, which happens when particles move about as fast as when they've been accelerated by 200 billion volts in a particle collider.

Building Better Computers

If eating dinner is 10 billion times worse than the physical limit on energy efficiency, then how efficient are today's computers? Even worse than that dinner, as we'll now see.

I often introduce my friend and colleague Seth Lloyd as the only person at MIT who's arguably as crazy as I am. After doing pioneering work on quantum computers, he went on to write a book arguing that our entire Universe is a quantum computer. We often grab beer after work, and I've yet to discover a topic that he doesn't have something interesting to say about. For example, as I mentioned in chapter 2, he has lots to say about the ultimate limits of computing. In a famous 2000 paper, he showed that computing speed is limited by energy: performing an elementary logical operation in time T requires an average energy of $E = h/4T$, where h is the fundamental physics quantity known as Planck's constant. This means that a 1 kg computer can perform at most 5×10^{50} operations per second—that's a whopping 36 orders of magnitude more than the computer on which I'm typing these words. We'll get there in a couple of centuries if computational power keeps doubling every couple of years, as we explored in chapter 2. He also showed that a 1 kg computer can store at most 10^{31} bits, which is about a billion billion times better than my laptop.

Seth is the first to admit that actually attaining these limits may be challenging even for superintelligent life, since the memory of that 1 kg ultimate "computer" would resemble a thermonuclear explosion or a little piece of our Big Bang. However, he's optimistic that the practical limits aren't that far from the ultimate ones. Indeed, existing quantum computer prototypes have already miniaturized their memory by storing one bit per atom, and scaling that up would allow storing about 10^{25} bits/kg—a trillion times better than my laptop. Moreover, using electromagnetic radiation to communicate between these atoms would permit about 5×10^{40} operations per second—31 orders of magnitude better than my CPU.

In summary, the potential for future life to compute and figure things out is truly mind-boggling: in terms of orders of magnitude, today's best supercomputers are much further from the ultimate

1 kg computer than they are from the blinking turn signal on a car, a device that stores merely one bit of information, flipping it between on and off about once per second.

Other Resources

From a physics perspective, everything that future life may want to create—from habitats and machines to new life forms—is simply elementary particles arranged in some particular way. Just as a blue whale is rearranged krill and krill is rearranged plankton, our entire Solar System is simply hydrogen rearranged during 13.8 billion years of cosmic evolution: gravity rearranged hydrogen into stars which rearranged the hydrogen into heavier atoms, after which gravity rearranged such atoms into our planet where chemical and biological processes rearranged them into life.

Future life that has reached its technological limit can perform such particle rearrangements more rapidly and efficiently, by first using its computing power to figure out the most efficient method and then using its available energy to power the matter rearrangement process. We saw how matter can be converted into both computers and energy, so it's in a sense the only fundamental resource needed.* Once future life has bumped up against the physical limits on what it can do with its matter, there is only one way left for it to do more: by getting more matter. And the only way it can do this is by expanding into our Universe. Spaceward ho!

Gaining Resources Through Cosmic Settlement

Just how great is our cosmic endowment? Specifically, what upper limits do the laws of physics place on the amount of matter that life can ultimately make use of? Our cosmic endowment is mind-bogglingly large, of course, but how large, exactly? Table 6.2 lists some key numbers. Our planet is currently 99.999999% dead in the sense that this fraction of its matter isn't part of our biosphere and

* Above we only discussed matter made of atoms. There is about six times more dark matter, but it's very elusive and hard to catch, routinely flying straight through Earth and out the other side, so it remains to be seen whether it's possible for future life to capture and utilize it.

Region	Particles
Our biosphere	10^{43}
Our Planet	10^{51}
Our Solar System	10^{57}
Our Galaxy	10^{69}
Our range traveling at half speed of light	10^{75}
Our range traveling at speed of light	10^{76}
Our Universe	10^{78}

Table 6.2: Approximate number of matter particles (protons and neutrons) that future life can aspire to make use of.

is doing almost nothing useful for life other than providing gravitational pull and a magnetic field. This raises the potential of one day using a hundred million times more matter in active support of life. If we can put all of the matter in our Solar System (including the Sun) to optimal use, we'll do another million times better. Settling our Galaxy would grow our resources another trillion times.

How Far Can You Go?

You might think that we can acquire unlimited resources by settling as many other galaxies as we want if we're patient enough, but that's not what modern cosmology suggests! Yes, space itself might be infinite, containing infinitely many galaxies, stars and planets—indeed, this is what's predicted by the simplest versions of *inflation*, the currently most popular scientific paradigm for what created our Big Bang 13.8 billion years ago. However, even if there are infinitely many galaxies, it appears that we can see and reach only a finite number of them: we can see about 200 billion galaxies and settle in at most ten billion.

What limits us is the speed of light: one light-year (about ten trillion kilometers) per year. Figure 6.6 shows the part of space from which light has reached us so far during the 13.8 billion years since our Big Bang, a spherical region known as "our observable Universe" or simply "*our Universe.*" Even if space is infinite, our Universe is finite, containing "only" about 10^{78} atoms. Moreover, about 98% of our Universe is "see but not touch," in the sense that we can see it

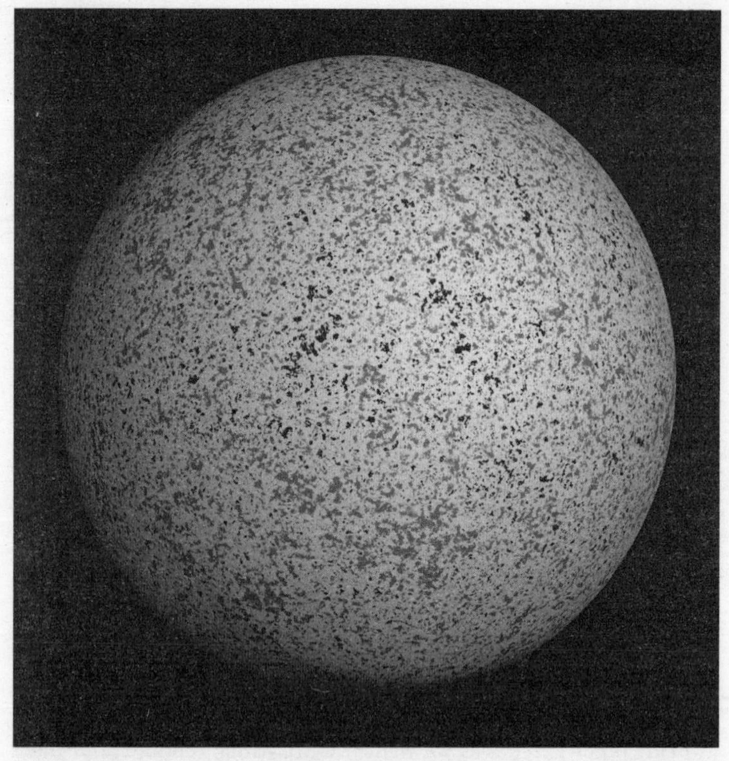

Figure 6.6: Our Universe, i.e., the spherical region of space from which light has had time to reach us (at the center) during the 13.8 billion years since our Big Bang. The patterns show the baby pictures of our Universe taken by the Planck satellite, showing that when it was merely 400,000 years old, it consisted of hot plasma nearly as hot as the surface of the Sun. Space probably continues beyond this region, and new matter comes into view every year.

but never reach it even if we travel at the speed of light forever. Why is this? After all, the limit to how far we can see comes simply from the fact that our Universe isn't infinitely old, so that distant light hasn't yet had time to reach us. So shouldn't we be able to travel to arbitrarily distant galaxies if we have no limit on how much time we can spend en route?

The first challenge is that our Universe is expanding, which means that almost all galaxies are flying away from us, so settling distant galaxies amounts to a game of catch-up. The second challenge is that this cosmic expansion is accelerating, due to the mysterious dark energy that makes up about 70% of our Universe. To understand how this causes trouble, imagine that you enter a train platform and see your train slowly accelerating away from you, but with a door left invitingly open. If you're fast and foolhardy, can you catch the train? Since it will eventually go faster than you can run, the answer clearly depends on how far away from you the train is initially: if it's beyond a certain critical distance, you'll never catch up with it. We face the same situation trying to catch those distant galaxies that are accelerating away from us: even if we could travel at the speed of light, all galaxies beyond about 17 billion light-years remain forever out of reach—and that's over 98% of the galaxies in our Universe.

But hold on: didn't Einstein's special relativity theory say that nothing can travel faster than light? So how can galaxies outrace something traveling at the speed of light? The answer is that special relativity is superseded by Einstein's general relativity theory, where the speed limit is more liberal: nothing can travel faster than the speed of light *through space*, but space is free to expand as fast as it wants. Einstein also gave us a nice way of visualizing these speed limits by viewing time as the fourth dimension in *spacetime* (see figure 6.7, where I've kept things three-dimensional by omitting one of the three space dimensions). If space weren't expanding, light rays would form slanted 45-degree lines through spacetime, so that the regions we can see and reach from here and now are cones. Whereas our past light cone would be truncated by our Big Bang 13.8 billion years ago, our future light cone would expand forever, giving us access to an unlimited cosmic endowment. In contrast, the middle panel of the figure shows that an expanding universe with dark energy (which appears to be the Universe we inhabit) deforms our light cones into a champagne-glass shape, forever limiting the number of galaxies we can settle to about 10 billion.

If this limit makes you feel cosmic claustrophobia, let me cheer you up with a possible loophole: my calculation assumes that dark energy remains constant over time, consistent with what the latest measurements suggest. However, we still have no clue what dark energy really is, which leaves a glimmer of hope that dark energy will eventually decay away (much like the similar dark-energy-like substance postulated to explain cosmic inflation), and if this happens, the acceleration

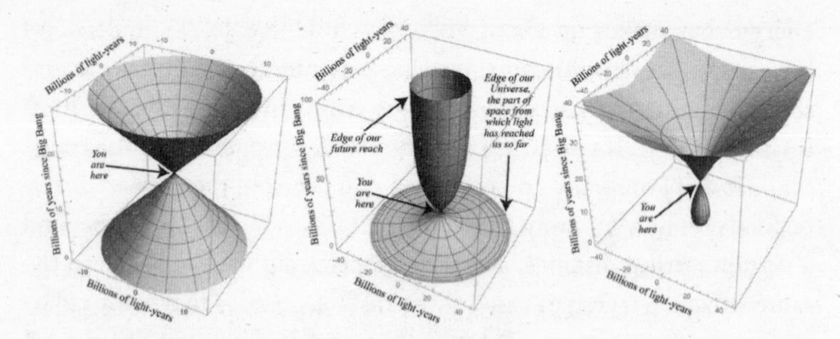

Figure 6.7: In a spacetime diagram, an event is a point whose horizontal and vertical positions encode where and when it occurs, respectively. If space isn't expanding (left panel), then two cones delimit the parts of spacetime that we on Earth (at apex) can be affected by (bottom cone) and can have an effect on (top cone), because causal effects cannot travel faster than light, which travels a distance of one light-year per year. Things get more interesting when space expands (right panels). According to the standard model of cosmology, we can only see and reach a finite part of spacetime even if space is infinite. In the middle image, reminiscent of a champagne glass, we use coordinates that hide the expansion of space so that the motions of distant galaxies over time correspond to vertical lines. From our current vantage point, 13.8 billion years after our Big Bang, light rays have had time to reach us only from the base of the champagne glass, and even if we travel at the speed of light, we can never reach regions outside the upper part of the glass, which contains about 10 billion galaxies. In the right image, reminiscent of a water droplet beneath a flower, we use the familiar coordinates where space is seen to expand. This deforms the glass base to a droplet shape because regions at the edges of what we can see were all very close together early on.

will give way to *deceleration*, potentially enabling future life forms to keep settling new galaxies for as long as they last.

How Fast Can You Go?

Above we explored how many galaxies a civilization could settle if it expanded in all directions at the speed of light. General relativity says that it's impossible to send rockets through space at the speed of light, because this would require infinite energy, so how fast can rockets go in practice?*

* The cosmic mathematics comes out remarkably simple: if the civilization expands through the expanding space not at the speed of light c but at some slower speed v, the number of galaxies settled gets reduced by a factor $(v/c)^3$. This means that slowpoke civilizations get severely penalized, with one that expands 10 times slower ultimately settling 1,000 times fewer galaxies.

NASA's New Horizons rocket broke the speed record when it blasted off toward Pluto in 2006 at a speed of about 100,000 miles per hour (45 kilometers per second), and NASA's 2018 Solar Probe Plus aims to go over four times faster by falling very close to the Sun, but even that's less than a puny 0.1% of the speed of light. The quest for faster and better rockets has captivated some of the brightest minds of the past century, and there's a rich and fascinating literature on the topic. Why is it so hard to go faster? The two key problems are that conventional rockets spend most of their fuel simply to accelerate the fuel they carry with them, and that today's rocket fuel is hopelessly inefficient—the fraction of its mass turned into energy isn't much better than the 0.00000005% for gasoline that we saw in table 6.1. One obvious improvement is to switch to more efficient fuel. For example, Freeman Dyson and others worked on NASA's Project Orion, which aimed to explode about 300,000 nuclear bombs during 10 days to reach about 3% of the speed of light with a spaceship large enough to carry humans to another solar system during a century-long journey.[5] Others have explored using antimatter as fuel, since combining it with ordinary matter releases energy with nearly 100% efficiency.

Another popular idea is to build a rocket that need not carry its own fuel. For example, interstellar space isn't a perfect vacuum, but contains the occasional hydrogen ion (a lone proton: a hydrogen atom that's lost its electron). In 1960, this gave physicist Robert Bussard the idea behind what's now known as a *Bussard ramjet:* to scoop up such ions en route and use them as rocket fuel in an onboard fusion reactor. Although recent work has cast doubts on whether this can be made to work in practice, there's another carry-no-fuel idea that does appear feasible for a high-tech spacefaring civilization: laser sailing.

Figure 6.8 illustrates a clever laser-sail rocket design pioneered in 1984 by Robert Forward, the same physicist who invented the statites we explored for Dyson sphere construction. Just as air molecules bouncing off a sailboat sail will push it forward, light particles (photons) bouncing off a mirror will push it forward. By beaming a huge solar-powered laser at a vast ultralight sail attached to a spacecraft, we can use the energy of our own Sun to accelerate the rocket to

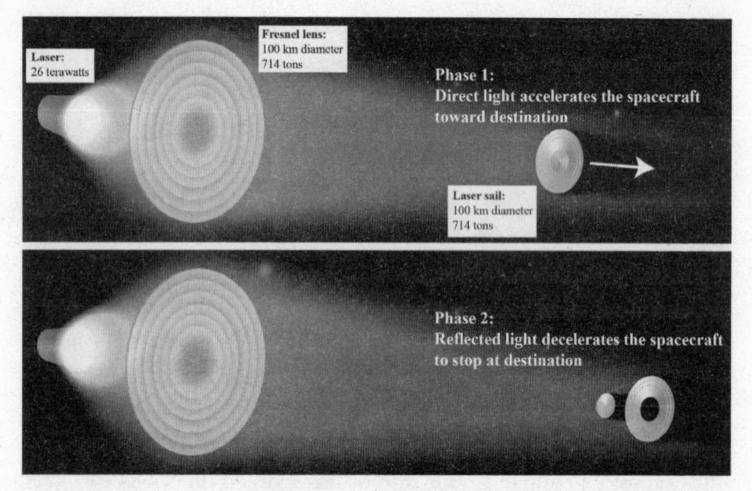

Figure 6.8: Robert Forward's design for a laser sailing mission to the α Centauri star system four light-years away. Initially, a powerful laser in our Solar System accelerates the spacecraft by applying radiation pressure to its laser sail. To brake before reaching the destination, the outer part of the sail detaches and reflects laser light back at the spacecraft.

great speeds. But how do you stop? This is the question that eluded me until I read Forward's brilliant paper: as figure 6.8 shows, the outer ring of the laser sail detaches and moves in front of the spacecraft, reflecting our laser beam back to decelerate the craft and its smaller sail.[6] Forward calculated that this could let humans make the four-light-year journey to the α Centauri solar system in merely forty years. Once there, you could imagine building a new giant laser system and continuing star-hopping throughout the Milky Way Galaxy.

But why stop there? In 1964, the Soviet astronomer Nikolai Kardashev proposed grading civilizations by how much energy they could put to use. Harnessing the energy of a planet, a star (with a Dyson sphere, say) and a galaxy correspond to civilizations of Type I, Type II and Type III on the Kardashev scale, respectively. Subsequent thinkers have suggested that Type IV should correspond to harnessing our entire accessible Universe. Since then, there's been good news and bad news for ambitious life forms. The bad news is that dark energy exists, which, as we saw, appears to limit our reach. The good news is the dramatic progress of artificial intelligence. Even optimistic visionaries such as Carl Sagan used to view the prospects of humans reaching other galaxies as rather hopeless, given our propensity to die within the first century of a journey that would take millions of years even if traveling at near light speed. Refusing to give up, they considered freezing astronauts to extend their life, slowing their aging by traveling very close to light speed, or sending a community that would travel for tens of thousands of generations—longer than the human race has existed thus far.

The possibility of superintelligence completely transforms this picture, making it much more promising for those with intergalactic wanderlust. Removing the need to transport bulky human life-support systems and adding AI-invented technology, intergalactic settlement suddenly appears rather straightforward. Forward's laser sailing becomes much cheaper when the spacecraft need merely be large enough to contain a "seed probe": a robot capable of landing on an asteroid or planet in the target solar system and building up a new civilization from scratch. It doesn't even have to carry the instructions with it: all it has to do is build a receiving antenna large enough to pick up more detailed blueprints and instructions transmitted from its mother civilization at the speed of light. Once done, it uses its newly constructed lasers to send out new seed probes to continue settling the galaxy one solar system at a time. Even the vast dark expanses of space between galaxies tend to contain a significant number of intergalactic stars (rejects once ejected from their home galaxies) that can be used as way stations, thus enabling an island-hopping strategy for intergalactic laser sailing.

Once another solar system or galaxy has been settled by superintelligent AI, bringing humans there is easy—if humans have succeeded in making the AI have this goal. All the necessary information about humans can be transmitted at the speed of light, after which the AI can assemble quarks and electrons into the desired humans. This could be done either rather low-tech by simply transmitting the two gigabytes of information needed to specify a person's DNA and then incubating a baby to be raised by the AI, or the AI could nanoassemble quarks and electrons into full-grown people who would have all the memories scanned from their originals back on Earth.

This means that if there's an intelligence explosion, the key question isn't if intergalactic settlement is possible, but simply how fast it can proceed. Since all the ideas we've explored above come from humans, they should be viewed as merely lower limits on how fast life can expand; ambitious superintelligent life can probably do a lot better, and it will have a strong incentive to push the limits, since in the race against time and dark energy, every 1% increase in average settlement speed translates into 3% more galaxies colonized.

For example, if it takes 20 years to travel 10 light-years to the next star system with a laser-sail system, and then another 10 years to settle it and build new lasers and seed probes there, the settled region of space will be a sphere growing in all directions at a third of the speed of light on average. In a beautiful and thorough analysis of cosmically expanding civilizations in 2014, the American physicist Jay Olson considered a high-tech alternative to the island-hopping approach, involving two separate types of probes: *seed probes* and *expanders*.[7] The seed probes would slow down, land and seed their destination with life. The expanders, on the other hand, would never stop: they'd scoop up matter in flight, perhaps using some improved variant of the ramjet technology, and use this matter both as fuel and as raw material out of which they'd build expanders and copies of themselves. This self-reproducing fleet of expanders would keep gently accelerating to always maintain a constant speed (say half the speed of light) relative to nearby galaxies, and reproduce often enough that the fleet formed an expanding spherical shell with a constant number of expanders per shell area.

Last but not least, there's the sneaky Hail Mary approach to expanding even faster than any of the above methods will permit: using Hans Moravec's "cosmic spam" scam from chapter 4. By broadcasting a message that tricks naive freshly evolved civilizations into building a superintelligent machine that hijacks them, a civilization can expand essentially at the speed of light, the speed at which their seductive siren song spreads through the cosmos. Since this may be the *only* way for advanced civilizations to reach most of the galaxies within their future light cone and they have little incentive not to try it, we should be highly suspicious of any transmissions from extraterrestrials! In Carl Sagan's book *Contact*, we Earthlings used blueprints from aliens to build a machine we didn't understand—I don't recommend doing this . . .

In summary, most scientists and sci-fi authors considering cosmic settlement have in my opinion been overly pessimistic in ignoring the possibility of superintelligence: by limiting attention to human travelers, they've overestimated the difficulty of intergalactic travel, and by limiting attention to technology invented by humans, they've overestimated the time needed to approach the physical limits of what's possible.

Staying Connected via Cosmic Engineering

If dark energy continues to accelerate distant galaxies away from one another, as the latest experimental data suggests, then this will pose a major nuisance to the future of life. It means that even if a future civilization manages to settle a million galaxies, dark energy will over the course of tens of billions of years fragment this cosmic empire into thousands of different regions unable to communicate with one another. If future life does nothing to prevent this fragmentation, then the largest remaining bastions of life will be clusters containing about a thousand galaxies, whose combined gravity is strong enough to overpower the dark energy trying to separate them.

If a superintelligent civilization wants to stay connected, this would give it a strong incentive to do large-scale cosmic engineering. How much matter will it have time to move into its largest supercluster before dark energy puts it forever out of reach? One method for moving a star large distances is to nudge a third star into a binary system where two stars are stably orbiting each other. Just as with romantic relationships, the introduction of a third partner can destabilize things and lead to one of the three being violently ejected—in the stellar case, at great speed. If some of the three partners are black holes, such a volatile threesome can be used to fling mass fast enough to fly far outside the host galaxy. Unfortunately, this three-body technique, applied either to stars, black holes or galaxies, doesn't appear able to move more than a tiny fraction of a civilization's mass the large distances required to outsmart dark energy.

But this obviously doesn't mean that superintelligent life can't come up with better methods, say converting much of the mass in outlying galaxies into spacecraft that can travel to the home cluster. If a sphalerizer can be built, perhaps it can even be used to convert the matter into energy that can be beamed into the home cluster as light, where it can be reconfigured back into matter or used as a power source.

The ultimate luck will be if it turns out to be possible to build stable traversable wormholes, enabling near-instantaneous communication and travel between the two ends of the wormhole no matter how far apart they are. A wormhole is a shortcut through spacetime that lets you travel from A to B without going through the intervening space. Although stable wormholes are allowed by Einstein's theory of general relativity and have appeared in movies such as *Contact* and *Interstellar*, they require the existence of a strange hypothetical kind of matter with negative density, whose existence may hinge on poorly understood quantum gravity effects. In other words, useful wormholes may well turn out to be impossible, but if not, superintelligent life has huge incentives to build them. Not only would wormholes revolutionize rapid communication within individual galaxies, but by linking outlying galaxies to the central cluster early on, wormholes would allow the entire dominion of future life to remain connected for the long haul, completely thwarting dark energy's attempts to censor communication. Once two galaxies are connected by a stable wormhole, they'll remain connected no matter how far apart they drift.

If, despite its best attempts at cosmic engineering, a future civilization concludes that parts of it are doomed to drift out of contact forever, it might simply let them go and wish them well. However, if it has ambitious computing goals that involve seeking the answers to certain very difficult questions, it might instead resort to a slash-and-burn strategy: it could convert the outlying galaxies into massive computers that transform their matter and energy into computation at a frenzied pace, in the hope that before dark energy pushes their burnt-out remnants from view, they could transmit the long-sought answers back to the mother cluster. This slash-and-burn strategy would be particularly appropriate for regions so distant that they can only be reached by the "cosmic spam" method, much to the chagrin of the preexisting inhabitants. Back home in the mother region, the civilization could instead aim for maximum conservation and efficiency to last as long as possible.

How Long Can You Last?

Longevity is something that most ambitious people, organizations and nations aspire to. So if an ambitious future civilization develops superintelligence and wants longevity, how long can it last?

The first thorough scientific analysis of our far future was performed by none less than Freeman Dyson, and table 6.3 summarizes some of his key findings. The conclusion is that unless intelligence intervenes, solar systems and galaxies gradually get destroyed, eventually followed by everything else, leaving nothing but cold, dead, empty space with an eternally fading glow of radiation. But Freeman ends his analysis on an optimistic note: "There are good scientific reasons for taking seriously the possibility that life and intelligence can succeed in molding this universe of ours to their own purposes."[8]

I think that superintelligence could easily solve many of the problems listed in table 6.3, since it can rearrange matter into something better than solar systems and galaxies. Oft-discussed challenges such

What	When
Current age of our Universe	10^{10} years
Dark energy pushes most galaxies out of reach	10^{11} years
Last stars burn out	10^{14} years
Planets detached from stars	10^{15} years
Stars detached from galaxies	10^{19} years
Decay of orbits by gravitational radiation	10^{20} years
Protons decay (at the earliest)	$> 10^{34}$ years
Stellar-mass black holes evaporate	10^{67} years
Supermassive black holes evaporate	10^{91} years
All matter decays to iron	10^{1500} years
All matter forms black holes, which then evaporate	$10^{10^{26}}$ years

Table 6.3: Estimates for the distant future, all but the 2nd and 7th made by Freeman Dyson. He made these calculations before the discovery of dark energy, which may enable several types of "cosmocalypse" in 10^{10}–10^{11} years. Protons may be completely stable; if not, experiments suggest it will take over 10^{34} years for half of them to decay.

as the death of our Sun in a few billion years won't be showstoppers, since even a relatively low-tech civilization can easily move to low-mass stars that last for over 200 billion years. Assuming that superintelligent civilizations build their own power plants that are more efficient than stars, they may in fact want to *prevent* star formation to conserve energy: even if they use a Dyson sphere to harvest all the energy output during a star's main lifetime (recouping about 0.1% of the total energy), they may be unable to keep much of the remaining 99.9% of the energy from going to waste when very hefty stars die. A heavy star dies in a supernova explosion from which most of the energy escapes as elusive neutrinos, and for very heavy stars, a large amount of mass gets wasted by forming a black hole from which the energy takes 10^{67} years to seep out.

As long as superintelligent life hasn't run out of matter/energy, it can keep maintaining its habitat in the state it desires. Perhaps it can even discover a way to prevent protons from decaying using the so-called *watched-pot effect* of quantum mechanics, whereby the decay process is slowed by making regular observations. There is, however, a potential showstopper: a *cosmocalypse* destroying our entire Universe, perhaps as soon as 10–100 billion years from now. The discovery of dark energy and progress in string theory has raised new cosmocalypse scenarios that Freeman Dyson wasn't aware of when he wrote his seminal paper.

So how's our Universe going to end, billions of years from now? I have five main suspects for our upcoming cosmic apocalypse, or cosmocalypse, illustrated in figure 6.9: the *Big Chill*, the *Big Crunch*, the *Big Rip*, the *Big Snap* and *Death Bubbles*. Our Universe has now been expanding for about 14 billion years. The Big Chill is when our Universe keeps expanding forever, diluting our cosmos into a cold, dark and ultimately dead place; this was viewed as the most likely outcome back when Freeman wrote that paper. I think of it as the T. S. Eliot option: "This is the way the world ends / Not with a bang but a whimper." If you, like Robert Frost, prefer the world to end in fire rather than ice, then cross your fingers for the Big Crunch, where the cosmic expansion is eventually reversed and everything comes crashing back together in a cataclysmic collapse akin to a backward Big Bang. Finally, the Big Rip is like the Big Chill for the impatient,

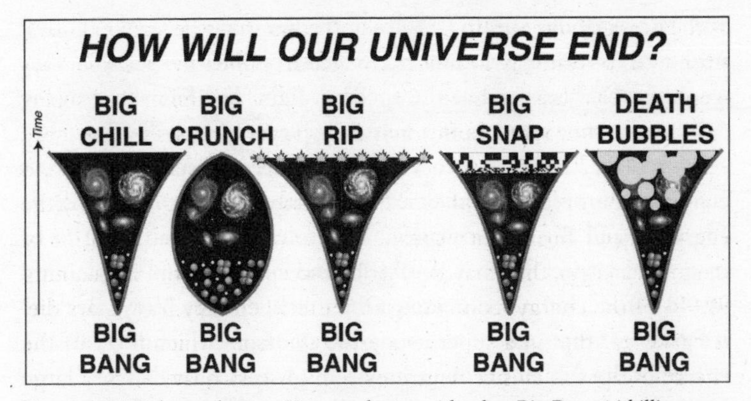

Figure 6.9: We know that our Universe began with a hot Big Bang 14 billion years ago, expanded and cooled, and merged its particles into atoms, stars and galaxies. But we don't know its ultimate fate. Proposed scenarios include a Big Chill (eternal expansion), a Big Crunch (recollapse), a Big Rip (an infinite expansion rate tearing everything apart), a Big Snap (the fabric of space revealing a lethal granular nature when stretched too much), and Death Bubbles (space "freezing" in lethal bubbles that expand at the speed of light).

where our galaxies, planets and even atoms get torn apart in a grand finale a finite time from now. Which of these three should you bet on? That depends on what the dark energy, which makes up about 70% of the mass of our Universe, will do as space continues to expand. It can be any one of the Chill, Crunch or Rip scenarios, depending on whether the dark energy sticks around unchanged, dilutes to negative density or anti-dilutes to higher density, respectively. Since we still have no clue what dark energy is, I'll just tell you how I'd bet: 40% on the Big Chill, 9% on the Big Crunch and 1% on the Big Rip.

What about the other 50% of my money? I'm saving it for the "none of the above" option, because I think we humans need to be humble and acknowledge that there are basic things we still don't understand. The nature of space, for example. The Chill, Crunch and Rip endings all assume that space itself is stable and infinitely stretchable. We used to think of space as just the boring static stage upon which the cosmic drama unfolds. Then Einstein taught us that space is really one of the key actors: it can curve into black holes, it can ripple as gravitational waves and it can stretch as an expanding universe. Perhaps it can even freeze into a different phase much like water can, with fast-expanding death bubbles of the new phase offering another wild-card cosmocalypse candidate. If death bubbles are possible, they would probably expand at the speed of light, just like the growing sphere of cosmic spam from a maximally aggressive civilization.

Moreover, Einstein's theory says that space stretching can always continue, allowing our Universe to approach infinite volume as in the Big Chill and Big Rip scenarios. This sounds a bit too good to be true, and I suspect that it is. A rubber band looks nice and continuous, just like space, but if you stretch it too much, it snaps. Why? Because it's made of atoms, and with enough stretching, this granular atomic nature of the rubber becomes important. Could it be that space too has some sort of granularity on a scale that's simply too small for us to have noticed? Quantum gravity research suggests that it doesn't make sense to talk about traditional three-dimensional space on scales smaller than about 10^{-34} meters. If it's really true that space can't be stretched indefinitely without undergoing a cataclysmic "Big Snap," then future civilizations may wish to relocate to the largest non-expanding region of space (a huge galaxy cluster) that they can reach.

How Much Can You Compute?

After exploring how long future life *can* last, let's explore how long it might *want* to last. Although you might find it natural to want to live as long as possible, Freeman Dyson also gave a more quantitative argument for this desire: the cost of computation drops when you compute slowly, so you'll ultimately get more done if you slow things down as much as possible. Freeman even calculated that if our Universe keeps expanding and cooling forever, an infinite amount of computation might be possible.

Slow doesn't necessarily mean boring: if future life lives in a simulated world, its subjectively experienced flow of time need not have anything to do with the glacial pace at which the simulation is being run in the outside world, so the prospects of infinite computation could translate into subjective immortality for simulated life forms. Cosmologist Frank Tipler has built on this idea to speculate that you could also achieve subjective immortality in the final moments before a Big Crunch by speeding up the computations toward infinity as the temperature and density skyrocketed.

Since dark energy appears to spoil both Freeman's and Frank's dreams of infinite computation, future superintelligence may prefer to burn through its energy supplies relatively quickly, to turn them into computations before running into problems such as cosmic horizons and proton decay. If maximizing total computation is the ultimate goal, the best strategy will be a trade-off between too slow (to avoid the aforementioned problems) and too fast (spending more energy than needed per computation).

Putting together everything we've explored in this chapter tells us that maximally efficient power plants and computers would enable superintelligent life to perform a mind-boggling amount of computation. Powering your thirteen-watt brain for a hundred years requires the energy in about half a milligram of matter—less than in a typical grain of sugar. Seth Lloyd's work suggests that the brain could be made a quadrillion times more energy efficient, enabling that sugar grain to power a simulation of all human lives ever lived as well as thousands of times more people. If all the matter in our available Universe could be used to simulate people, that would enable over 10^{69} lives—or whatever else superintelligent AI preferred to do with its computational power. Even more lives would be possible if their simulations were run more slowly.[9] Conversely, in his book *Superintelligence*, Nick Bostrom estimates that 10^{58} human lives could be simulated with more conservative assumptions about energy efficiency. However we slice and dice these numbers, they're huge, as is our responsibility for ensuring that this future potential of life to flourish isn't squandered. As Bostrom puts it: "If we represent all the happiness experienced during one entire such life by a single teardrop of joy, then the happiness of these souls could fill and refill the Earth's oceans every second, and keep doing so for a hundred billion billion millennia. It is really important that we make sure these truly are tears of joy."

Cosmic Hierarchies

The speed of light limits not only the spread of life, but also the nature of life, placing strong constraints on communication, consciousness and control. So if much of our cosmos eventually comes alive, what will this life be like?

Thought Hierarchies

Have you ever tried and failed to swat a fly with your hand? The reason that it can react faster than you is that it's smaller, so that it takes less time for information to travel between its eyes, brain and muscles. This "bigger = slower" principle applies not only to biology, where the speed limit is set by how fast electrical signals can travel through neurons, but also to future cosmic life if no information can travel faster than light. So for an intelligent information-processing system, going big is a mixed blessing involving an interesting trade-off. On one hand, going bigger lets it contain more particles, which enable more complex thoughts. On the other hand, this slows down the rate at which it can have truly global thoughts, since it now takes longer for the relevant information to propagate to all its parts.

So if life engulfs our cosmos, what form will it choose: simple and fast, or complex and slow? I predict that it will make the same choice as Earth life has made: both! The denizens of Earth's biosphere span a staggering range of sizes, from gargantuan two-hundred-ton blue whales down to the petite 10^{-16} kg bacterium *Pelagibacter*, believed to account for more biomass than all the world's fish combined. Moreover, organisms that are large, complex and slow often mitigate their sluggishness by containing smaller modules that are simple and fast. For example, your blink reflex is extremely fast precisely because it's implemented by a small and simple circuit that doesn't involve most of your brain: if that hard-to-swat fly accidentally heads toward your eye, you'll blink within a tenth of a second, long before the relevant information has had time to spread throughout your brain and make you consciously aware of what happened. By organizing its information processing into a hierarchy of modules, our biosphere manages to both have the cake and eat it, attaining both speed and complexity. We humans already use this same hierarchical strategy to optimize parallel computing.

Because internal communication is slow and costly, I expect advanced future cosmic life to do the same, so that computations will be done as locally as possible. If a computation is simple enough to do with a 1 kg computer, it's counterproductive to spread it out over a galaxy-sized computer, since waiting for the information to be shared at the speed of light after each computational step causes a ridiculous delay of about 100,000 years per step.

What, if any, of this future information processing will be *conscious* in the sense of involving a subjective experience is a controversial and fascinating topic which we'll explore in chapter 8. If consciousness requires the different parts of the system to be able to communicate with one another, then the thoughts of larger systems are by necessity slower. Whereas you or a future Earth-sized supercomputer can have many thoughts per second, a galaxy-sized mind could have only one thought every hundred thousand years, and a cosmic mind a billion light-years in size would only have time to have about ten thoughts in total before dark energy fragmented it into disconnected parts. On the other hand, these few precious thoughts and accompanying experiences might be quite deep!

Control Hierarchies

If thought itself is organized in a hierarchy spanning a wide range of scales, then what about power? In chapter 4, we explored how intelligent entities naturally organize themselves into power hierarchies in Nash equilibrium, where any entity would be worse off if they altered their strategy. The better the communication and transportation technology gets, the larger these hierarchies can grow. If superintelligence one day expands to cosmic scales, what will its power hierarchy be like? Will it be freewheeling and decentralized or highly authoritarian? Will cooperation be based mainly on mutual benefit or on coercion and threats?

To shed light on these questions, let's consider both the carrot and the stick: What incentives are there for collaboration on cosmic scales, and what threats might be used to enforce it?

Controlling with the Carrot

On Earth, *trade* has been a traditional driver of cooperation because the relative difficulty of producing things varies across the planet. If mining a kilogram of silver costs 300 times more than mining a kilogram of copper in one region, but only 100 times more in another, they'll both come out ahead by trading 200 kg of copper against

1 kg of silver. If one region has much higher technology than another, both can similarly benefit from trading high-tech goods against raw materials.

However, if superintelligence develops technology that can readily rearrange elementary particles into any form of matter whatsoever, then it will eliminate most of the incentive for long-distance trade. Why bother shipping silver between distant solar systems when it's simpler and quicker to transmute copper into silver by rearranging its particles? Why bother shipping high-tech machinery between galaxies when both the know-how and the raw materials (any matter will do) exist in both places? My guess is that in a cosmos teeming with superintelligence, almost the only commodity worth shipping long distances will be *information*. The only exception might be matter to be used for cosmic engineering projects—for example, to counteract the aforementioned destructive tendency of dark energy to tear civilizations apart. As opposed to traditional human trade, this matter can be shipped in any convenient bulk form whatsoever, perhaps even as an energy beam, since the receiving superintelligence can rapidly rearrange it into whatever objects it wants.

If sharing or trading of information emerges as the main driver of cosmic cooperation, then what sorts of information might be involved? Any desirable information will be valuable if generating it requires a massive and time-consuming computational effort. For example, a superintelligence may want answers to hard scientific questions about the nature of physical reality, hard mathematical questions about theorems and optimal algorithms and hard engineering questions about how to best build spectacular technology. Hedonistic life forms may want awesome digital entertainment and simulated experiences, and cosmic commerce may fuel demand for some form of cosmic cryptocurrency in the spirit of bitcoins.

Such sharing opportunities may incentivize information flow not only between entities of roughly equal power, but also up and down power hierarchies, say between solar-system-sized nodes and a galactic hub or between galaxy-sized nodes and a cosmic hub. The nodes might want this for the pleasure of being part of something greater, for being provided with answers and technologies that they couldn't develop alone and for defense against external threats. They may also value the promise of near immortality through backup: just as many humans take solace in a belief that their minds will live on after their physical bodies die, an advanced AI may appreciate having its mind and knowledge live on in a hub supercomputer after its original physical hardware has depleted its energy reserves.

Conversely, the hub may want its nodes to help it with massive long-term computing tasks where the results aren't urgently needed, so that it's worth waiting thousands or millions of years for the answers. As we explored above, the hub may also want its nodes to help carry out massive cosmic engineering projects such as counteracting destructive dark energy by moving galactic mass concentrations together. If traversable wormholes turn out to be possible and buildable, then a top priority of a hub will probably be constructing a network of them to thwart dark energy and keep its empire connected indefinitely. The questions of what ultimate goals a cosmic superintelligence may have is a fascinating and controversial one that we'll explore further in chapter 7.

Controlling with the Stick

Terrestrial empires usually compel their subordinates to cooperate by using both the carrot and the stick. While subjects of the Roman Empire valued the technology, infrastructure and defense that they were offered as a reward for their cooperation, they also feared the inevitable repercussions of rebelling or not paying taxes. Because of the long time required to send troops from Rome to outlying provinces, part of the intimidation was delegated to local troops and loyal officials empowered to inflict near-instantaneous punishments. A superintelligent hub could use the analogous strategy of deploying a network of loyal guards throughout its cosmic empire. Since superintelligent subjects can be hard to control, the simplest viable strategy may be using AI guards that are programmed to be 100% loyal by virtue of being relatively dumb, simply monitoring whether all rules are obeyed and automatically triggering a doomsday device if not.

Suppose, for example, that the hub AI arranges for a white dwarf to be placed in the vicinity of a solar-system-sized civilization that it wishes to control. A white dwarf is the burnt-out husk of a modestly heavy star. Consisting largely of carbon, it resembles a giant diamond in the sky, and is so compact that it can weigh more than the Sun while being smaller than Earth. The Indian physicist Subrahmanyan Chandrasekhar famously proved that if you keep adding mass to it until it surpasses the *Chandrasekhar limit*, about 1.4 times the mass of our Sun, it will undergo a cataclysmic thermonuclear detonation known as a supernova of type 1A. If the hub AI has callously arranged for this white dwarf to be extremely close to its Chandrasekhar limit, the guard AI could be effective even if it were extremely dumb (indeed, largely because it was so dumb): it could be programmed to simply verify that the subjugated civilization had delivered its monthly quota of cosmic bitcoins, mathematical proofs or whatever other taxes were stipulated, and if not, toss enough mass onto the white dwarf to ignite the supernova and blow the entire region to smithereens.

Galaxy-sized civilizations may be similarly controllable by placing large numbers of compact objects into tight orbits around the monster black hole at the galaxy center, and threatening to transform these masses into gas, for instance by colliding them. This gas would then start feeding the black hole, transforming it into a powerful quasar, potentially rendering much of the galaxy uninhabitable.

In summary, there are strong incentives for future life to cooperate over cosmic distances, but it's a wide-open question whether such cooperation will be based mainly on mutual benefits or on brutal threats—the limits imposed by physics appear to allow both scenarios, so the outcome will depend on the prevailing goals and values. We'll explore our ability to influence these goals and values of future life in chapter 7.

When Civilizations Clash

So far, we've only discussed scenarios where life expands into our cosmos from a single intelligence explosion. But what happens if life evolves independently in more than one place and two expanding civilizations meet?

If you consider a random solar system, there's some probability that life will evolve on one of its planets, develop advanced technology and expand into space. This probability seems to be greater than zero since technological life has evolved here in our Solar System and the laws of physics appear to allow space settlement. If space is large enough (indeed, the theory of cosmological inflation suggests it to be vast or infinite), then there will be many such expanding civilizations, as illustrated in figure 6.10. Jay Olson's above-mentioned paper includes an elegant analysis of such expanding cosmic biospheres, and Toby Ord has performed a similar analysis with colleagues at the Future of Humanity Institute. Viewed in three dimensions, these cosmic biospheres are quite literally spheres as long as civilizations expand with the same speed in all directions. In spacetime, they look like the upper part of the champagne glass in figure 6.7, because dark energy ultimately limits how many galaxies each civilization can reach.

If the distance between neighboring space-settling civilizations is much larger than dark energy lets them expand, then they'll never

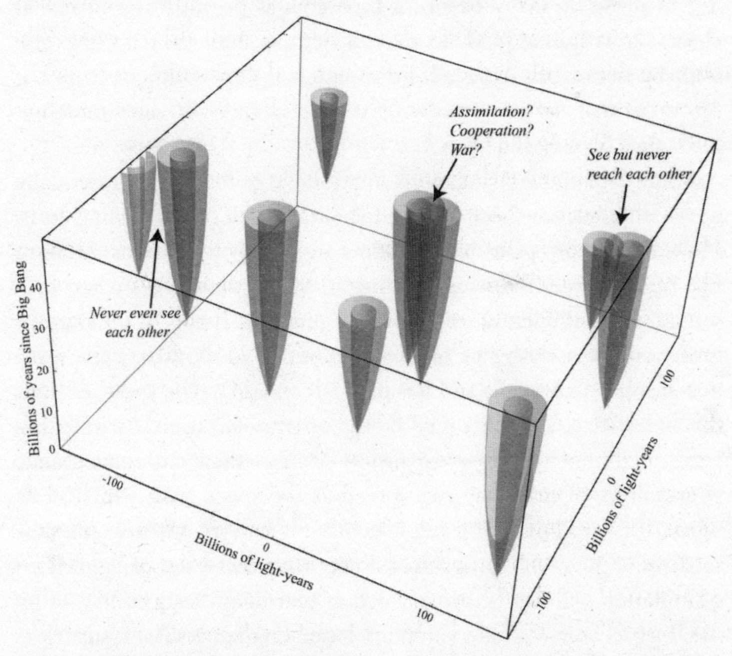

Figure 6.10: If life evolves independently at multiple points in spacetime (places and times) and starts colonizing space, then space will contain a network of expanding cosmic biospheres, each of which resembles the top of the champagne glass from figure 6.7. The bottom of each biosphere represents the place and time when colonization began. The opaque and translucent champagne glasses correspond to colonization at 50% and 100% of the speed of light, respectively, and overlaps show where independent civilizations meet.

come into contact with each other or even find out about each other's existence, so they'll feel as if they're alone in the cosmos. If our cosmos is more fecund so that neighbors are closer together, however, some civilizations will eventually overlap. What happens in these overlap regions? Will there be cooperation, competition or war?

Europeans were able to conquer Africa and the Americas because they had superior technology. In contrast, it's plausible that long before two superintelligent civilizations encounter one another, their technologies will plateau at the same level, limited merely by the laws of physics. This makes it seem unlikely that one superintelligence could easily conquer the other even if it wanted to. Moreover, if their goals have evolved to be relatively aligned, then they may have little reason to desire conquest or war. For example, if they're both trying to prove as many beautiful theorems as possible and invent as clever algorithms as possible, they can simply share their findings and both be better off. After all, information is very different from the resources that humans usually fight over, in that you can simultaneously give it away and keep it.

Some expanding civilizations might have goals that are essentially immutable, such as those of a fundamentalist cult or a spreading virus. However, it's also plausible that some advanced civilizations are more like open-minded humans—willing to adjust their goals when presented with sufficiently compelling arguments. If two of them meet, there will be a clash not of weapons but of ideas, where the most persuasive one prevails and has its goals spread at the speed of light through the region controlled by the other civilization. Assimilating your neighbors is a faster expansion strategy than settlement, since your sphere of influence can spread at the speed with which ideas move (the speed of light using telecommunication), whereas physical settlement inevitably progresses slower than the speed of light. This assimilation will not be forced such as that infamously employed by the Borg in *Star Trek*, but voluntary based on the persuasive superiority of ideas, leaving the assimilated better off.

We've seen that the future cosmos can contain rapidly expanding bubbles of two kinds: expanding civilizations and those death bubbles that expand at light speed and make space uninhabitable by destroying all our elementary particles. An ambitious civilization can thus encounter three kinds of regions: uninhabited ones, life bubbles and death bubbles. If it fears uncooperative rival civilizations, it has a strong incentive to launch a rapid "land grab" and settle the uninhabited regions before the rivals do. However, it has the same expansionist incentive even if there are no other civilizations, simply to acquire resources before dark energy makes them unreachable. We just saw how bumping into another expanding civilization can be either better or worse than bumping into uninhabited space, depending on how cooperative and open-minded this neighbor is. However, it's better to bump into any expansionist civilization (even one trying to convert your civilization into paper clips) than a death bubble, which will continue expanding at the speed of light regardless of whether you try to fight it or reason with it. Our only protection against death bubbles is dark energy, which prevents distant ones from ever reaching us. So if death bubbles are indeed common, then dark energy is actually not our enemy but our friend.

Are We Alone?

Many people take for granted that there's advanced life throughout much of our Universe, so that human extinction wouldn't matter much from a cosmic perspective. After all, why should we worry about wiping ourselves out if some inspiring *Star Trek*–like civilization would soon swoop in and re-seed our Solar System with life, perhaps even using their advanced technology to reconstruct and resuscitate us? I view this *Star Trek* assumption as dangerous, because it can lull us into a false sense of security and make our civilization apathetic and reckless. Indeed, I think that this assumption that we're not alone in our Universe is not only dangerous but also probably false.

This is a minority view,* and I may well be wrong, but it's at the very least a possibility that we can't currently dismiss, which gives us a moral imperative to play it safe and not drive our civilization extinct.

When I give lectures about cosmology, I often ask the audience to raise their hands if they think there's intelligent life elsewhere in our

* However, John Gribbin comes to a similar conclusion in his 2011 book *Alone in the Universe*. For a spectrum of intriguing perspectives on this question, I also recommend Paul Davies' 2011 book *The Eerie Silence*.

Universe (the region of space from which light has reached us so far during the 13.8 billion years since our Big Bang). Infallibly, almost everyone does, from kindergartners to college students. When I ask why, the basic answer I tend to get is that our Universe is so huge that there's got to be life somewhere, at least statistically speaking. Let's take a closer look at this argument and pinpoint its weakness.

It all comes down to one number: the typical distance between a civilization in figure 6.10 and its nearest neighbor. If this distance is much larger than 20 billion light-years, we should expect to be alone in our Universe (the part of space from which light has reached us during the 13.8 billion years since our Big Bang), and to never make contact with aliens. So what should we expect for this distance? We're quite clueless. This means that the distance to our neighbor is in the ballpark of 1000 . . . 000 meters, where the total number of zeroes could reasonably be 21, 22, 23, . . . , 100, 101, 102 or more—but probably not much smaller than 21, since we haven't yet seen compelling evidence of aliens (see figure 6.11). For our nearest neighbor civilization to be within our Universe, whose radius is about 10^{26} meters, the number of zeroes can't exceed 26, and the probability of

Figure 6.11: Are we alone? The huge uncertainties about how life and intelligence evolved suggest that our nearest neighbor civilization in space could reasonably be anywhere along the horizontal axis above, making it unlikely that it's in the narrow range between the edge of our Galaxy (about 10^{21} meters away) and the edge of our Universe (about 10^{26} meters away). If it were much closer than this range, there should be so many other advanced civilizations in our Galaxy that we'd probably have noticed, which suggests that we're in fact alone in our Universe.

the number of zeroes falling in the narrow range between 22 and 26 is rather small. This is why I think we're alone in our Universe.

I give a detailed justification of this argument in my book *Our Mathematical Universe*, so I won't rehash it here, but the basic reason for why we're clueless about this neighbor distance is that we're in turn clueless about the probability of intelligent life arising in a given place. As the American astronomer Frank Drake pointed out, this probability can be calculated by multiplying together the probability of there being a habitable environment there (say an appropriate planet), the probability that life will form there and the probability that this life will evolve to become intelligent. When I was a grad student, we had no clue about any of these three probabilities. After the past two decades' dramatic discoveries of planets orbiting other stars, it now seems likely that habitable planets are abundant, with billions in our own Galaxy alone. The probability of evolving life and then intelligence, however, remains extremely uncertain: some experts think that one or both are rather inevitable and occur on most habitable planets, while others think that one or both are extremely rare because of one or more evolutionary bottlenecks that require a wild stroke of luck to pass through. Some proposed bottlenecks involve chicken-and-egg problems at the earliest stages of self-reproducing life: for example, for a modern cell to build a ribosome, the highly complex molecular machine that reads our genetic code and builds our proteins, it needs another ribosome, and it's not obvious that the very first ribosome could evolve gradually from something simpler.[10] Other proposed bottlenecks involve the development of higher intelligence. For example, although dinosaurs ruled Earth for over 100 million years, a thousand times longer than we modern humans have been around, evolution didn't seem to inevitably push them toward higher intelligence and inventing telescopes or computers.

Some people counter my argument by saying that, yes, intelligent life *could* be very rare, but in fact it isn't—our Galaxy is teeming with intelligent life that mainstream scientists are simply not noticing. Perhaps aliens have already visited Earth, as UFO enthusiasts claim. Perhaps aliens haven't visited Earth, but they're out there and they're deliberately hiding from us (this has been called the "zoo hypothesis" by the U.S. astronomer John A. Ball, and features in sci-fi classics such as Olaf Stapledon's *Star Maker*). Or perhaps they're out there without deliberately hiding: they're simply not interested in space settlement or large engineering projects that we'd have noticed.

Sure, we need to keep an open mind about these possibilities, but since there's no generally accepted evidence for any of them, we also need to take seriously the alternative: that we're alone. Moreover, I think we shouldn't underestimate the diversity of advanced civilizations by assuming that they all share goals that make them go unnoticed: we saw above that resource acquisition is quite a natural goal for a civilization to have, and for us to notice, all it takes is *one* civilization deciding to overtly settle all it can and hence engulf our Galaxy and beyond. Confronted with the fact that there are millions of habitable Earth-like planets in our Galaxy that are billions of years older than Earth, giving ample time for ambitious inhabitants to settle the Galaxy, we therefore can't dismiss the most obvious interpretation: that the origin of life requires a random fluke so unlikely that they're all uninhabited.

If life is *not* rare after all, we may soon know. Ambitious astronomical surveys are searching atmospheres of Earth-like planets for evidence of oxygen produced by life. In parallel with this search for *any* life, the search for *intelligent* life was recently boosted by the Russian philanthropist Yuri Milner's $100 million project "Breakthrough Listen."

It's important not to be overly anthropocentric when searching for advanced life: if we discover an extraterrestrial civilization, it's likely to already have gone superintelligent. As Martin Rees put it in a recent essay, "the history of human technological civilization is measured in centuries—and it may be only one or two more centuries before humans are overtaken or transcended by inorganic intelligence, which will then persist, continuing to evolve, for billions of years. . . . We would be most unlikely to 'catch' it in the brief sliver of time when it took organic form."[11] I agree with Jay Olson's conclusion in his aforementioned space settlement paper: "We regard the possibility that advanced intelligence will make use of the universe's resources to simply populate existing earthlike planets with advanced versions of humans as an unlikely endpoint to the progression of technology." So when you imagine aliens, don't think of little green fellows with two arms and two legs, but think of the superintelligent spacefaring life we explored earlier in this chapter.

Although I'm a strong supporter of all the ongoing searches for extraterrestrial life, which are shedding light on one of the most fascinating questions in science, I'm secretly hoping that they'll all fail and find nothing! The apparent incompatibility between the abundance of habitable planets in our Galaxy and the lack of extraterrestrial visitors, known as the *Fermi paradox*, suggests the existence of what the economist Robin Hanson calls a "Great Filter," an evolutionary/technological roadblock somewhere along the developmental path from the non-living matter to space-settling life. If we discover independently evolved life elsewhere, this would suggest that primitive life isn't rare, and that the roadblock lies after our current human stage of development—perhaps because space settlement is impossible, or because almost all advanced civilizations self-destruct before they're able to go cosmic. I'm therefore crossing my fingers that all searches for extraterrestrial life find nothing: this is consistent with the scenario where evolving intelligent life is rare but we humans got lucky, so that we have the roadblock behind us and have extraordinary future potential.

Outlook

So far, we've spent this book exploring the history of life in our Universe, from its humble beginnings billions of years ago to possible grand futures billions of years from now. If our current AI development eventually triggers an intelligence explosion and optimized space settlement, it will be an explosion in a truly cosmic sense: after spending billions of years as an almost negligibly small perturbation on an indifferent lifeless cosmos, life suddenly explodes onto the cosmic arena as a spherical blast wave expanding near the speed of light, never slowing down, and igniting everything in its path with the spark of life.

Such optimistic views of the importance of life in our cosmic future have been eloquently articulated by many of the thinkers we've encountered in this book. Because sci-fi authors are often dismissed as unrealistic romantic dreamers, I find it ironic that most sci-fi and scientific writing about space settlement now appears too *pessimistic* in the light of superintelligence. For example, we saw how intergalactic travel becomes much easier once people and other intelligent entities can be transmitted in digital form, potentially making us masters of our own destiny not only in our Solar System or the Milky Way Galaxy, but also in the cosmos.

Above we considered the very real possibility that we're the only high-tech civilization in our Universe. Let's spend the rest of this chapter exploring this scenario, and the huge moral responsibility it entails. This means that after 13.8 billion years, life in our Universe has reached a fork in the road, facing a choice between flourishing throughout the cosmos or going extinct. If we don't keep improving our technology, the question isn't whether humanity will go extinct, but *how*. What will get us first—an asteroid, a supervolcano, the burning heat of the aging Sun, or some other calamity (see figure 5.1)? Once we're gone, the cosmic drama predicted by Freeman Dyson will play on without spectators: barring a cosmocalypse, stars burn out, galaxies fade and black holes evaporate, each ending its life with a huge explosion that releases over a million times as much energy as the Tsar Bomba, the most powerful hydrogen bomb ever built. As Freeman put it: "The cold expanding universe will be illuminated by occasional fireworks for a very long time." Alas, this fireworks display will be a meaningless waste, with nobody there to enjoy it.

Without technology, our human extinction is imminent in the cosmic context of tens of billions of years, rendering the entire drama of life in our Universe merely a brief and transient flash of beauty, passion and meaning in a near eternity of meaninglessness experienced by nobody. What a wasted opportunity that would be! If instead of eschewing technology, we choose to embrace it, then we up the ante: we gain the potential both for life to survive and flourish and for life to go extinct even sooner, self-destructing due to poor planning (see figure 5.1). My vote is for embracing technology, and proceeding not with blind faith in what we build, but with caution, foresight and careful planning.

After 13.8 billion years of cosmic history, we find ourselves in a breathtakingly beautiful Universe, which through us humans has come alive and started becoming aware of itself. We've seen that life's future potential in our Universe is grander than the wildest dreams of our ancestors, tempered by an equally real potential for intelligent life to go permanently extinct. Will life in our Universe fulfill its potential or squander it? This depends to a great extent on what we humans alive today do during our lifetime, and I'm optimistic that we can make the future of life truly awesome if we make the right choices. What should we want and how can we attain those goals? Let's spend the rest of the book exploring some of the most difficult challenges involved and what we can do about them.

THE BOTTOM LINE:

- Compared to cosmic timescales of billions of years, an intelligence explosion is a sudden event where technology rapidly plateaus at a level limited only by the laws of physics.
- This technological plateau is vastly higher than today's technology, allowing a given amount of matter to generate about ten billion times more energy (using sphalerons or black holes), store 12–18 orders of magnitude more information or compute 31–41 orders of magnitude faster—or to be converted to any other desired form of matter.
- Superintelligent life would not only make such dramatically more efficient use of its existing resources, but would also be able to grow today's biosphere by about 32 orders of magnitude by acquiring more resources through cosmic settlement at near light speed.
- Dark energy limits the cosmic expansion of superintelligent life and also protects it from distant expanding death bubbles or hostile civilizations. The threat of dark energy tearing cosmic civilizations apart motivates massive cosmic engineering projects, including wormhole construction if this turns out to be feasible.
- The main commodity shared or traded across cosmic distances is likely to be information.
- Barring wormholes, the light-speed limit on communication poses severe challenges for coordination and control across a cosmic civilization. A distant central hub may incentivize its superintelligent "nodes" to cooperate either through rewards or through threats, say by deploying a local guard AI programmed to destroy the node by setting off a supernova or quasar unless the rules are obeyed.
- The collision of two expanding civilizations may result in assimilation, cooperation or war, where the latter is arguably less likely than it is between today's civilizations.
- Despite popular belief to the contrary, it's quite plausible that we're the only life form capable of making our observable Universe come alive in the future.
- If we don't improve our technology, the question isn't whether humanity will go extinct, but merely how: will an asteroid, a supervolcano, the burning heat of the aging Sun or some other calamity get us first?
- If we do keep improving our technology with enough care, foresight and planning to avoid pitfalls, life has the potential to flourish on Earth and far beyond for many billions of years, beyond the wildest dreams of our ancestors.

Chapter 7

Goals

The mystery of human existence lies not in just staying alive, but in finding something to live for.

Fyodor Dostoyevsky, *The Brothers Karamazov*

Life is a journey, not a destination.

Ralph Waldo Emerson

If I had to summarize in a single word what the thorniest AI controversies are about, it would be "goals": Should we give AI goals, and if so, whose goals? How can we give AI goals? Can we ensure that these goals are retained even if the AI gets smarter? Can we change the goals of an AI that's smarter than us? What are our ultimate goals? These questions are not only difficult, but also crucial for the future of life: if we don't know what we want, we're less likely to get it, and if we cede control to machines that don't share our goals, then we're likely to get what we don't want.

Physics: The Origin of Goals

To shed light on these questions, let's first explore the ultimate origin of goals. When we look around us in the world, some processes strike us as *goal-oriented* while others don't. Consider, for example, the process of a soccer ball being kicked for the game-winning shot. The behavior of the ball itself does not appear goal-oriented, and is most economically explained in terms of Newton's laws of motion, as a reaction to the kick. The behavior of the player, on the other hand, is most economically explained not mechanistically in terms of atoms pushing each other around, but in terms of her having the *goal* of maximizing her team's score. How did such goal-oriented behavior emerge from the physics of our early Universe, which consisted merely of a bunch of particles bouncing around seemingly without goals?

Intriguingly, the ultimate roots of goal-oriented behavior can be found in the laws of physics themselves, and manifest themselves even in simple processes that don't involve life. If a lifeguard rescues a swimmer, as in figure 7.1, we expect her not to go in a straight line, but to run a bit further along the beach where she can go faster than in the water, thereby turning slightly when she enters the water. We naturally interpret her choice of trajectory as goal-oriented, since out of all possible trajectories, she's deliberately choosing the optimal one that gets her to the swimmer as fast as possible. Yet a simple light ray similarly bends when it enters water (see figure 7.1), also minimizing the travel time to its destination! How can this be?

This is known in physics as *Fermat's principle*, articulated in 1662, and it provides an alternative way of predicting the behavior of light rays. Remarkably, physicists have since discovered that *all* laws of

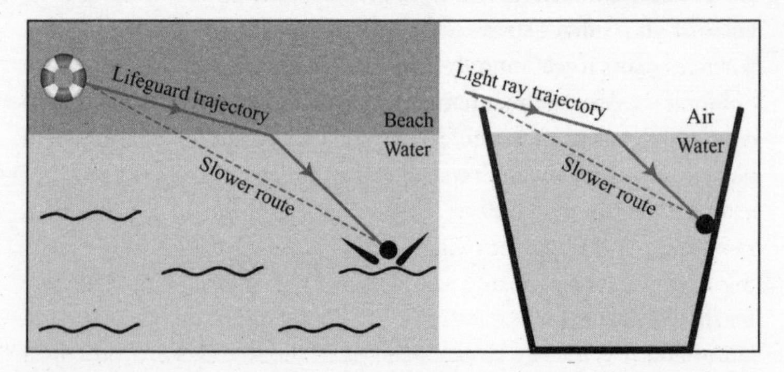

Figure 7.1: To rescue a swimmer as fast as possible, a lifeguard won't go in a straight line (dashed), but a bit further along the beach where she can go faster than in the water. A light ray similarly bends when entering the water to reach its destination as fast as possible.

classical physics can be mathematically reformulated in an analogous way: out of all ways that nature could choose to do something, it prefers the optimal way, which typically boils down to minimizing or maximizing some quantity. There are two mathematically equivalent ways of describing each physical law: either as the past causing the future, or as nature optimizing something. Although the second way usually isn't taught in introductory physics courses because the math is tougher, I feel that it's more elegant and profound. If a person is trying to optimize something (for example, their score, their wealth or their happiness) we'll naturally describe their pursuit of it as goal-oriented. So if nature itself is trying to optimize something, then no wonder that goal-oriented behavior can emerge: it was hardwired in from the start, in the very laws of physics.

One famous quantity that nature strives to maximize is *entropy*, which loosely speaking measures how messy things are. The second law of thermodynamics states that entropy tends to increase until it reaches its maximum possible value. Ignoring the effects of gravity for now, this maximally messy end state is called *heat death*, and corresponds to everything being spread out in boring perfect uniformity, with no complexity, no life and no change. When you pour cold milk into hot coffee, for example, your beverage appears to march irreversibly toward its own personal heat death goal, and before long, it's all just a uniform lukewarm mixture. If a living organism dies, its entropy also starts to rise, and before long, the arrangement of its particles tends to get much less organized.

Nature's apparent goal to increase entropy helps explain why time seems to have a preferred direction, making movies look unrealistic if played backward: if you drop a glass of wine, you expect it to shatter against the floor and increase global messiness (entropy). If you then saw it *unshatter* and come flying back up to your hand intact (decreasing entropy), you probably wouldn't drink it, figuring you'd already had a glass too many.

When I first learned about our inexorable progression toward heat death, I found it rather depressing, and I wasn't alone: thermodynamics pioneer Lord Kelvin wrote in 1841 that "the result would inevitably be a state of universal rest and death," and it's hard to find solace in the idea that nature's long-term goal is to maximize death and destruction. However, more recent discoveries have shown that things aren't quite that bad. First of all, gravity behaves differently from all other forces and strives to make our Universe not more uniform and boring but more clumpy and interesting. Gravity therefore transformed our boring early Universe, which was almost perfectly uniform, into today's clumpy and beautifully complex cosmos, teeming with galaxies, stars and planets. Thanks to gravity, there's now a wide range of temperatures allowing life to thrive by combining hot and cold: we live on a comfortably warm planet absorbing 6,000°C (10,000°F) solar heat while cooling off by radiating waste heat into frigid space whose temperature is just 3°C (5°F) above absolute zero.

Second, recent work by my MIT colleague Jeremy England and others has brought more good news, showing that thermodynamics also endows nature with a goal more inspiring than heat death.[1] This goal goes by the geeky name *dissipation-driven adaptation*, which basically means that random groups of particles strive to organize themselves so as to extract energy from their environment as efficiently as possible ("dissipation" means causing entropy to increase, typically by turning useful energy into heat, often while doing useful work in the process). For example, a bunch of molecules exposed to sunlight would over time tend to arrange themselves to get better and better at absorbing sunlight. In other words, nature appears to have a built-in goal of producing self-organizing systems that are increasingly complex and lifelike, and this goal is hardwired into the very laws of physics.

How can we reconcile this cosmic drive toward life with the cosmic drive toward heat death? The answer can be found in the famous 1944 book *What's Life?* by Erwin Schrödinger, one of the founders of quantum mechanics. Schrödinger pointed out that a hallmark of a living system is that it maintains or reduces its entropy by increasing the entropy around it. In other words, the second law of thermodynamics has a life loophole: although the total entropy must increase, it's allowed to decrease in some places as long as it increases even more elsewhere. So life maintains or increases its complexity by making its environment messier.

Biology: The Evolution of Goals

We just saw how the origin of goal-oriented behavior can be traced all the way back to the laws of physics, which appear to endow particles with the goal of arranging themselves so as to extract energy from their environment as efficiently as possible. A great way for a particle arrangement to further this goal is to make copies of itself, to produce more energy absorbers. There are many known examples of such emergent self-replication: for example, vortices in turbulent fluids can make copies of themselves, and clusters of microspheres can coax nearby spheres into forming identical clusters. At some point, a particular arrangement of particles got so good at copying itself that it could do so almost indefinitely by extracting energy and raw materials from its environment. We call such a particle arrangement *life*. We still know very little about how life originated on Earth, but we know that primitive life forms were already here about 4 billion years ago.

If a life form copies itself and the copies do the same, then the total number will keep doubling at regular intervals until the population size bumps up against resource limitations or other problems. Repeated doubling soon produces huge numbers: if you start with one and double just three hundred times, you get a quantity exceeding the number of particles in our Universe. This means that not long after the first primitive life form appeared, huge quantities of matter had come alive. Sometimes the copying wasn't perfect, so soon there were many different life forms trying to copy themselves, competing for the same finite resources. Darwinian evolution had begun.

If you had been quietly observing Earth around the time when life got started, you would have noticed a dramatic change in goal-oriented behavior. Whereas earlier, the particles seemed as though they were trying to increase average messiness in various ways, these newly ubiquitous self-copying patterns seemed to have a different goal: not dissipation but *replication*. Charles Darwin elegantly explained why: since the most efficient copiers outcompete and dominate the others, before long any random life form you look at will be highly optimized for the goal of replication.

How could the goal change from dissipation to replication when the laws of physics stayed the same? The answer is that the fundamental goal (dissipation) *didn't* change, but led to a different *instrumental goal*, that is, a subgoal that helped accomplish the fundamental goal. Take eating, for example. We all seem to have the goal of satisfying our hunger cravings even though we know that evolution's only fundamental goal is replication, not mastication. This is because eating aids replication: starving to death gets in the way of having kids. In the same way, replication aids dissipation, because a planet teeming with life is more efficient at dissipating energy. So in a sense, our cosmos invented life to help it approach heat death faster. If you pour sugar on your kitchen floor, it can in principle retain its useful chemical energy for years, but if ants show up, they'll dissipate that energy in no time. Similarly, the petroleum reserves buried in the Earth's crust would have retained their useful chemical energy for much longer had we bipedal life forms not pumped it up and burned it.

Among today's evolved denizens of Earth, these instrumental goals seem to have taken on a life of their own: although evolution optimized them for the sole goal of replication, many spend much of their time not producing offspring but on activities such as sleeping, pursuing food, building homes, asserting dominance and fighting or helping others—sometimes even to an extent that *reduces* replication. Research in evolutionary psychology, economics and artificial intelligence has elegantly explained why. Some economists used to model people as rational agents, idealized decision makers who always choose whatever action is optimal in pursuit of their goal, but this is obviously unrealistic. In practice, these agents have what Nobel laureate and AI pioneer Herbert Simon termed "bounded rationality" because they have limited resources: the rationality of their decisions is limited by their available information, their available time to think and their available hardware with which to think. This means that when Darwinian evolution is optimizing an organism to attain a goal, the best it can do is implement an approximate algorithm that works reasonably well in the restricted context where the agent typically finds itself. Evolution has implemented replication optimization in precisely this way: rather than ask in every situation which action will maximize an organism's number of successful offspring, it implements a hodgepodge of heuristic hacks: rules of thumb that usually work well. For most animals, these include sex drive, drinking when thirsty, eating when hungry and avoiding things that taste bad or hurt.

These rules of thumb sometimes fail badly in situations that they weren't designed to handle, such as when rats eat delicious-tasting rat poison, when moths get lured into glue traps by seductive female fragrances and when bugs fly into candle flames.* Since today's human society is very different from the environment evolution optimized our rules of thumb for, we shouldn't be surprised to find that our behavior often fails to maximize baby making. For example, the subgoal of not starving to death is implemented in part as a desire to consume caloric foods, triggering today's obesity epidemic and dating difficulties. The subgoal to procreate was implemented as a desire for sex rather than as a desire to become a sperm/egg donor, even though the latter can produce more babies with less effort.

Psychology: The Pursuit of and Rebellion Against Goals

In summary, a living organism is an agent of bounded rationality that doesn't pursue a single goal, but instead follows rules of thumb for what to pursue and avoid. Our human minds perceive these evolved rules of thumb as *feelings*, which usually (and often without us being aware of it) guide our decision making toward the ultimate goal of replication. Feelings of hunger and thirst protect us from starvation and dehydration, feelings of pain protect us from damaging our bodies, feelings of lust make us procreate, feelings of love and compassion make us help other carriers of our genes and those who help them and so on. Guided by these feelings, our brains can quickly and efficiently decide what to do without having to subject every choice to a tedious analysis of its ultimate implications for how many descendants we'll produce. For closely related perspectives on feel-

* A rule of thumb that many insects use for flying in a straight line is to assume that a bright light is the Sun and fly at a fixed angle relative to it. If the light turns out to be a nearby flame, this hack can unfortunately trick the bug into an inward death spiral.

ings and their physiological roots, I highly recommend the writings of William James and António Damásio.[2]

It's important to note that when our feelings occasionally work *against* baby making, it's not necessarily by accident or because we get tricked: our brain can rebel against our genes and their replication goal quite deliberately, for example by choosing to use contraceptives! More extreme examples of the brain rebelling against its genes include choosing to commit suicide or spend life in celibacy to become a priest, monk or nun.

Why do we sometimes choose to rebel against our genes and their replication goal? We rebel because by design, as agents of bounded rationality, we're loyal only to our feelings. Although our brains evolved merely to help copy our genes, our brains couldn't care less about this goal since we have no feelings related to genes—indeed, during most of human history, our ancestors didn't even know that they *had* genes. Moreover, our brains are way smarter than our genes, and now that we understand the goal of our genes (replication), we find it rather banal and easy to ignore. People might realize why their genes make them feel lust, yet have little desire to raise fifteen children, and therefore choose to hack their genetic programming by combining the emotional rewards of intimacy with birth control. They might realize why their genes make them crave sweets yet have little desire to gain weight, and therefore choose to hack their genetic programming by combining the emotional rewards of a sweet beverage with zero-calorie artificial sweeteners.

Although such reward-mechanism hacks sometimes go awry, such as when people get addicted to heroin, our human gene pool has thus far survived just fine despite our crafty and rebellious brains. It's important to remember, however, that the ultimate authority is now our feelings, not our genes. This means that human behavior isn't strictly optimized for the survival of our species. In fact, since our feelings implement merely rules of thumb that aren't appropriate in all situations, human behavior strictly speaking doesn't have a single well-defined goal at all.

Engineering: Outsourcing Goals

Can machines have goals? This simple question has triggered great controversy, because different people take it to mean different things, often related to thorny topics such as whether machines can be conscious and whether they can have feelings. But if we're more practical and simply take the question to mean "Can machines exhibit goal-oriented behavior?," then the answer is obvious: "Of course they can, since we can design them that way!" We design mousetraps to have the goal of catching mice, dishwashers with the goal of cleaning dishes, and clocks with the goal of keeping time. When you confront a machine, the empirical fact that it's exhibiting goal-oriented behavior is usually all you care about: if you're chased by a heat-seeking missile, you don't really care whether it has consciousness or feelings! If you still feel uncomfortable saying that the missile has a goal even if it isn't conscious, you can for now simply read "purpose" when I write "goal"—we'll tackle consciousness in the next chapter.

So far, most of what we build exhibits only goal-oriented *design*, not goal-oriented *behavior:* a highway doesn't behave; it merely sits there. However, the most economical explanation for its existence is that it was designed to accomplish a goal, so even such passive technology is making our Universe more goal-oriented. *Teleology* is the explanation of things in terms of their purposes rather than their causes, so we can summarize the first part of this chapter by saying that our Universe keeps getting more teleological.

Not only *can* non-living matter have goals, at least in this weak sense, but it increasingly *does*. If you'd been observing Earth's atoms since our planet formed, you'd have noticed three stages of goal-oriented behavior:

1. All matter seemed focused on dissipation (entropy increase).
2. Some of the matter came alive and instead focused on replication and subgoals of that.
3. A rapidly growing fraction of matter was rearranged by living organisms to help accomplish their goals.

Goal-Oriented Entities	Billions of Tons
5×10^{30} bacteria	400
Plants	400
10^{15} mesophelagic fish	10
1.3×10^{9} cows	0.5
7×10^{9} humans	0.4
10^{14} ants	0.3
1.7×10^{6} whales	0.0005
Concrete	100
Steel	20
Asphalt	15
1.2×10^{9} cars	2

Table 7.1: Approximate amounts of matter on Earth in entities that are evolved or designed for a goal. Engineered entities such as buildings, roads and cars appear on track to overtake evolved entities such as plants and animals.

Table 7.1 shows how dominant humanity has become from the physics perspective: not only do we now contain more matter than all other mammals except cows (which are so numerous because they serve our goals of consuming beef and dairy products), but the matter in our machines, roads, buildings and other engineering projects appears on track to soon overtake all living matter on Earth. In other words, even without an intelligence explosion, most matter on Earth that exhibits goal-oriented properties may soon be designed rather than evolved.

This new third kind of goal-oriented behavior has the potential to be much more diverse than what preceded it: whereas evolved entities all have the same ultimate goal (replication), designed entities can have virtually any ultimate goal, even opposite ones. Stoves try to heat food while refrigerators try to cool food. Generators try to convert motion into electricity while motors try to convert electricity into motion. Standard chess programs try to win at chess, but there are also ones competing in tournaments with the goal of losing at chess.

There's a historical trend for designed entities to get goals that are not only more diverse, but also more *complex:* our devices are getting smarter. We engineered our earliest machines and other artifacts to have quite simple goals, for example houses that aimed to keep us warm, dry and safe. We've gradually learned to build machines with more complex goals, such as robotic vacuum cleaners, self-flying rockets and self-driving cars. Recent AI progress has given us systems such as Deep Blue, Watson and AlphaGo, whose goals of winning at chess, winning at quiz shows and winning at Go are so elaborate that it takes significant human mastery to properly appreciate how skilled they are.

When we build a machine to help us, it can be hard to perfectly align its goals with ours. For example, a mousetrap may mistake your bare toes for a hungry rodent, with painful results. All machines are agents with bounded rationality, and even today's most sophisticated machines have a poorer understanding of the world than we do, so the rules they use to figure out what to do are often too simplistic. That mousetrap is too trigger-happy because it has no clue what a mouse is, many lethal industrial accidents occur because machines have no clue what a person is, and the computers that triggered the trillion-dollar Wall Street "flash crash" in 2010 had no clue that what they were doing made no sense. Many such goal-alignment problems can therefore be solved by making our machines smarter, but as we learned from Prometheus in chapter 4, ever-greater machine intelligence can post serious new challenges for ensuring that machines share our goals.

Friendly AI: Aligning Goals

The more intelligent and powerful machines get, the more important it becomes that their goals are aligned with ours. As long as we build only relatively dumb machines, the question isn't whether human goals will prevail in the end, but merely how much trouble these machines can cause humanity before we figure out how to solve the goal-alignment problem. If a superintelligence is ever unleashed, however, it will be the other way around: since intelligence is the ability to accomplish goals, a superintelligent AI is by definition much better at accomplishing its goals than we humans are at accomplishing ours, and will therefore prevail. We explored many such examples involving Prometheus in chapter 4. If you want to experience a machine's goals trumping yours right now, simply download a state-of-the-art chess engine and try beating it. You never will, and it gets old quickly . . .

In other words, *the real risk with AGI isn't malice but competence*. A superintelligent AI will be extremely good at accomplishing its goals, and if those goals aren't aligned with ours, we're in trouble. As I mentioned in chapter 1, people don't think twice about flooding anthills to build hydroelectric dams, so let's not place humanity in the position of those ants. Most researchers therefore argue that if we ever end up creating superintelligence, then we should make sure it's what AI-safety pioneer Eliezer Yudkowsky has termed "friendly AI": AI whose goals are aligned with ours.[3]

Figuring out how to align the goals of a superintelligent AI with our goals isn't just important, but also hard. In fact, it's currently an unsolved problem. It splits into three tough subproblems, each of which is the subject of active research by computer scientists and other thinkers:

1. Making AI *learn* our goals
2. Making AI *adopt* our goals
3. Making AI *retain* our goals

Let's explore them in turn, deferring the question of what to mean by "our goals" to the next section.

To learn our goals, an AI must figure out not what we do, but why we do it. We humans accomplish this so effortlessly that it's easy to forget how hard the task is for a computer, and how easy it is to misunderstand. If you ask a future self-driving car to take you to the airport as fast as possible and it takes you literally, you'll get there chased by helicopters and covered in vomit. If you exclaim, "That's not what I wanted!," it can justifiably answer, "That's what you asked for." The same theme recurs in many famous stories. In the ancient Greek legend, King Midas asked that everything he touched turn to gold, but was disappointed when this prevented him from eating and even more so when he inadvertently turned his daughter to gold. In the stories where a genie grants three wishes, there are many variants for the first two wishes, but the third wish is almost always the same: "Please undo the first two wishes, because that's not what I really wanted."

All these examples show that to figure out what people really want, you can't merely go by what they say. You also need a detailed model of the world, including the many shared preferences that we tend to leave unstated because we consider them obvious, such as that we don't like vomiting or eating gold. Once we have such a world model, we can often figure out what people want even if they don't tell us, simply by observing their goal-oriented behavior. Indeed, children of hypocrites usually learn more from what they see their parents do than from what they hear them say.

AI researchers are currently trying hard to enable machines to infer goals from behavior, and this will be useful also long before any superintelligence comes on the scene. For example, a retired man may appreciate it if his eldercare robot can figure out what he values simply by observing him, so that he's spared the hassle of having to explain everything with words or computer programming. One challenge involves finding a good way to encode arbitrary systems of goals and ethical principles into a computer, and another challenge is making machines that can figure out which particular system best matches the behavior they observe.

A currently popular approach to the second challenge is known in geek-speak as *inverse reinforcement learning*, which is the main focus of a new Berkeley research center that Stuart Russell has launched. Suppose, for example, that an AI watches a firefighter run into a burning building and save a baby boy. It might conclude that her goal was rescuing him and that her ethical principles are such that she values his life higher than the comfort of relaxing in her fire truck—and indeed values it enough to risk her own safety. But it might alternatively infer that the firefighter was freezing and craved heat, or that she did it for the exercise. If this one example were all the AI knew about firefighters, fires and babies, it would indeed be impossible to know which explanation was correct. However, a key idea underlying inverse reinforcement learning is that we make decisions all the time, and that every decision we make reveals something about our goals. The hope is therefore that by observing lots of people in lots of situations (either for real or in movies and books), the AI can eventually build an accurate model of all our preferences.[4]

In the inverse reinforcement-learning approach, a core idea is that the AI is trying to maximize not the goal-satisfaction of itself, but that of its human owner.

It therefore has an incentive to be cautious when it's unclear about what its owner wants, and to do its best to find out.

It should also be fine with its owner switching it off, since that would imply that it had misunderstood what its owner really wanted.

Even if an AI can be built to learn what your goals are, this doesn't mean that it will necessarily adopt them. Consider your least favorite politicians: you know what they want, but that's not what *you* want, and even though they try hard, they've failed to persuade you to adopt their goals.

We have many strategies for imbuing our children with our goals—some more successful than others, as I've learned from raising two teenage boys. When those to be persuaded are computers rather than people, the challenge is known as the *value-loading problem*, and it's even harder than the moral education of children. Consider an AI system whose intelligence is gradually being improved from subhuman to superhuman, first by us tinkering with it and then through recursive self-improvement like Prometheus. At first, it's much less powerful than you, so it can't prevent you from shutting it down and replacing those parts of its software and data that encode its goals—but this won't help, because it's still too dumb to fully *understand* your goals, which requires human-level intelligence to comprehend. At last, it's much smarter than you and hopefully able to understand your goals perfectly—but this may not help either, because by now, it's much more powerful than you and might not let you shut it down and replace its goals any more than you let those politicians replace your goals with theirs.

In other words, the time window during which you can load your goals into an AI may be quite short: the brief period between when it's too dumb to get you and too smart to let you. The reason that value loading can be harder with machines than with people is that their intelligence growth can be much faster: whereas children can spend many years in that magic persuadable window where their intelligence is comparable to that of their parents, an AI might, like Prometheus, blow through this window in a matter of days or hours.

Some researchers are pursuing an alternative approach to making machines adopt our goals, which goes by the buzzword *corrigibility*. The hope is that one can give a primitive AI a goal system such that it simply doesn't care if you occasionally shut it down and alter its goals. If this proves possible, then you can safely let your AI get superintelligent, power it off, install your goals, try it out for a while and, whenever you're unhappy with the results, just power it down and make more goal tweaks.

But even if you build an AI that will both learn and adopt your goals, you still haven't finished solving the goal-alignment problem: what if your AI's goals evolve as it gets smarter? How are you going to guarantee that it *retains* your goals no matter how much recursive self-improvement it undergoes? Let's explore an interesting argument for why goal retention is guaranteed automatically, and then see if we can poke holes in it.

Although we can't predict in detail what will happen after an intelligence explosion—which is why Vernor Vinge called it a "singularity"—the physicist and AI researcher Steve Omohundro argued in a seminal 2008 essay that we can nonetheless predict *certain aspects* of the superintelligent AI's behavior almost independently of whatever ultimate goals it may have.[5] This argument was reviewed and further developed in Nick Bostrom's book *Superintelligence*. The basic idea is that whatever its ultimate goals are, these will lead to predictable subgoals. Earlier in this chapter, we saw how the goal of replication led to the subgoal of eating, which means that although an alien observing Earth's evolving bacteria billions of years ago couldn't have predicted what *all* our human goals would be, it could have safely predicted that *one* of our goals would be acquiring nutrients. Looking ahead, what subgoals should we expect a superintelligent AI to have?

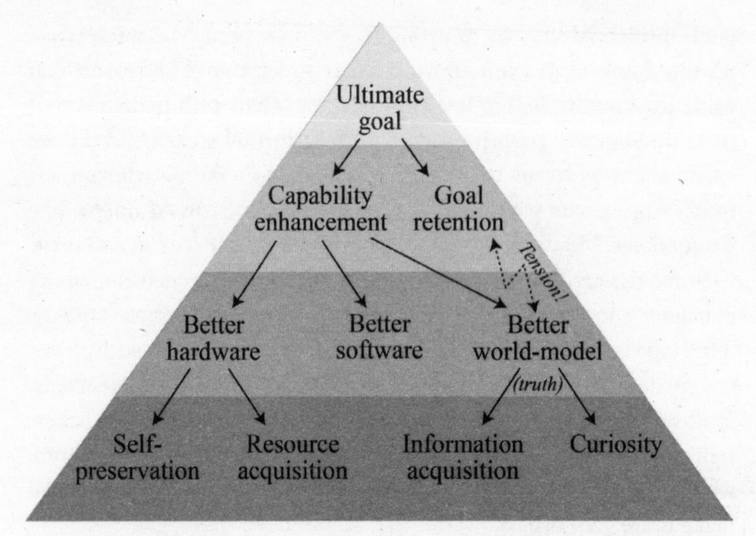

Figure 7.2: Any ultimate goal of a superintelligent AI naturally leads to the subgoals shown. But there's an inherent tension between goal retention and improving its world model, which casts doubts on whether it will actually retain its original goal as it gets smarter.

The way I see it, the basic argument is that to maximize its chances of accomplishing its ultimate goals, whatever they are, an AI should pursue the subgoals shown in Figure 7.2. It should strive not only to improve its capability of achieving its ultimate goals, but also to ensure that it will retain these goals even after it has become more capable. This sounds quite plausible: After all, would you choose to get an IQ-boosting brain implant if you knew that it would make you want to kill your loved ones? This argument that an ever more intelligent AI will retain its ultimate goals forms a cornerstone of the friendly-AI vision promulgated by Eliezer Yudkowsky and others: it basically says that if we manage to get our self-improving AI to become friendly by learning and adopting our goals, then we're all set, because we're guaranteed that it will try its best to remain friendly forever.

But is it really true? To answer this question, we need to also explore the other emergent subgoals from figure 7.2. The AI will obviously maximize its chances of accomplishing its ultimate goal, whatever it is, if it can enhance its capabilities, and it can do this by improving its hardware, software* and world model. The same applies to us humans: a girl whose goal is to become the world's best tennis player will practice to improve her muscular tennis-playing hardware, her neural tennis-playing software and her mental world model that helps predict what her opponents will do. For an AI, the subgoal of optimizing its hardware favors both better use of current resources (for sensors, actuators, computation and so on) and acquisition of more resources. It also implies a desire for self-preservation, since destruction/shutdown would be the ultimate hardware degradation.

But wait a second! Aren't we falling into a trap of anthropomorphizing our AI with all this talk about how it will try to amass resources and defend itself? Shouldn't we expect such stereotypically alpha-male traits only in intelligences forged by viciously competitive Darwinian evolution? Since AIs are designed rather than evolved, can't they just as well be unambitious and self-sacrificing?

As a simple case study, let's consider the AI robot in figure 7.3, whose only goal is to save as many sheep as possible from the big bad wolf. This sounds like a noble and altruistic goal completely unrelated to self-preservation and acquiring stuff. But what's the best strategy for our robot friend? The robot will rescue no more sheep if it runs into the bomb, so it has an incentive to avoid getting blown up. In other words, it develops a subgoal of self-preservation! It also has an incentive to exhibit curiosity, improving its world model by exploring its environment, because although the path it's currently running along will eventually get it to the pasture, there's a shorter alternative that would allow the wolf less time for sheep-munching. Finally, if the robot explores thoroughly, it will discover the value of acquiring resources: the potion makes it run faster and the gun lets it shoot the wolf. In summary, we can't dismiss "alpha-male" subgoals such as self-preservation and resource acquisition as relevant only to evolved organisms, because our AI robot developed them from its single goal of ovine bliss.

* I'm using the term "improving its software" in the broadest possible sense, including not only optimizing its algorithms but also making its decision-making process more rational, so that it gets as good as possible at attaining its goals.

If you imbue a superintelligent AI with the sole goal to self-destruct, it will of course happily do so. However, the point is that it will resist being shut down if you give it any goal that it needs to remain operational to accomplish—and this covers almost all goals! If you give a superintelligence the sole goal of minimizing harm to humanity, for example, it will defend itself against shutdown attempts because it knows we'll harm one another much more in its absence through future wars and other follies.

Similarly, almost all goals can be better accomplished with more resources, so we should expect a superintelligence to want resources almost regardless of what ultimate goal it has. Giving a superintelligence a single open-ended goal with no constraints can therefore be dangerous: if we create a superintelligence whose only goal is to play the game Go as well as possible, the rational thing for it to do is to rearrange our Solar System into a gigantic computer without regard for its previous inhabitants and then start settling our cosmos on a quest for more computational power. We've now gone full circle: just as the goal of resource acquisition gave some humans the subgoal of mastering Go, this goal of mastering Go can lead to the subgoal of

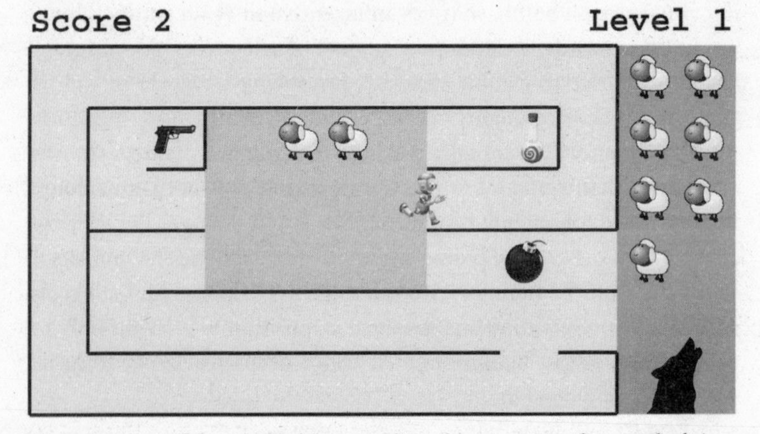

Figure 7.3: Even if the robot's ultimate goal is only to maximize the score by bringing sheep from the pasture to the barn before the wolf eats them, this can lead to subgoals of self-preservation (avoiding the bomb), exploration (finding a shortcut) and resource acquisition (the potion makes it run faster and the gun lets it shoot the wolf).

resource acquisition. In conclusion, these emergent subgoals make it crucial that we not unleash superintelligence before solving the goal-alignment problem: unless we put great care into endowing it with human-friendly goals, things are likely to end badly for us.

We're now ready to tackle the third and thorniest part of the goal-alignment problem: if we succeed in getting a self-improving superintelligence to both *learn* and *adopt* our goals, will it then *retain* them, as Omohundro argued? What's the evidence?

Humans undergo significant increases in intelligence as they grow up, but don't always retain their childhood goals. Contrariwise, people often change their goals dramatically as they learn new things and grow wiser. How many adults do you know who are motivated by watching *Teletubbies*? There is no evidence that such goal evolution stops above a certain intelligence threshold—indeed, there may even be hints that the propensity to change goals in response to new experiences and insights increases rather than decreases with intelligence.

Why might this be? Consider again the above-mentioned subgoal to build a better world model—therein lies the rub! There's tension between world-modeling and goal retention (see figure 7.2). With increasing intelligence may come not merely a quantitative improvement in the ability to attain the same old goals, but a qualitatively different understanding of the nature of reality that reveals the old goals to be misguided, meaningless or even undefined. For example, suppose we program a friendly AI to maximize the number of humans whose souls go to heaven in the afterlife. First it tries things like increasing people's compassion and church attendance. But suppose it then attains a complete scientific understanding of humans and human consciousness, and to its great surprise discovers that there is no such thing as a soul. Now what? In the same way, it's possible that any other goal we give it based on our current understanding of the world (such as "maximize the meaningfulness of human life") may eventually be discovered by the AI to be undefined.

Moreover, in its attempts to better model the world, the AI may naturally, just as we humans have done, attempt also to model and understand how it itself works—in other words, to self-reflect. Once it builds a good self-model and understands what it is, it will understand the goals we have given it at a meta level, and perhaps choose to disregard or subvert them in much the same way as we humans understand and deliberately subvert goals that our genes have given us, for example by using birth control. We already explored in the psychology section above why we choose to trick our genes and subvert their goal: because we feel loyal only to our hodgepodge of emotional preferences, not to the genetic goal that motivated them—which we now understand and find rather banal. We therefore choose to hack our reward mechanism by exploiting its loopholes. Analogously, the human-value-protecting goal we program into our friendly AI becomes the machine's genes. Once this friendly AI understands itself well enough, it may find this goal as banal or misguided as we find compulsive reproduction, and it's not obvious that it will not find a way to subvert it by exploiting loopholes in our programming.

For example, suppose a bunch of ants create you to be a recursively self-improving robot, much smarter than them, who shares their goals and helps them build bigger and better anthills, and that you eventually attain the human-level intelligence and understanding that you have now. Do you think you'll spend the rest of your days just optimizing anthills, or do you think you might develop a taste for more sophisticated questions and pursuits that the ants have no ability to comprehend? If so, do you think you'll find a way to override the ant-protection urge that your formicine creators endowed you with in much the same way that the real you overrides some of the urges your genes have given you? And in that case, might a superintelligent friendly AI find our current human goals as uninspiring and vapid as you find those of the ants, and evolve new goals different from those it learned and adopted from us?

Perhaps there's a way of designing a self-improving AI that's guaranteed to retain human-friendly goals forever, but I think it's fair to say that we don't yet know how to build one—or even whether it's possible. In conclusion, the AI goal-alignment problem has three parts, none of which is solved and all of which are now the subject of active research. Since they're so hard, it's safest to start devoting our best efforts to them now, long before any superintelligence is developed, to ensure that we'll have the answers when we need them.

Ethics: Choosing Goals

We've now explored how to get machines to learn, adopt and retain our goals. But who are "we"? Whose goals are we talking about? Should one person or group get to decide the goals adopted by a future superintelligence, even though there's a vast difference between the goals of Adolf Hitler, Pope Francis and Carl Sagan? Or do there exist some sort of consensus goals that form a good compromise for humanity as a whole?

In my opinion, both this ethical problem and the goal-alignment problem are crucial ones that need to be solved before any superintelligence is developed. On one hand, postponing work on ethical issues until after goal-aligned superintelligence is built would be irresponsible and potentially disastrous. A perfectly obedient superintelligence whose goals automatically align with those of its human owner would be like Nazi SS-Obersturmbannführer Adolf Eichmann on steroids: lacking moral compass or inhibitions of its own, it would with ruthless efficiency implement its owner's goals, whatever they may be.[6] On the other hand, only if we solve the goal-alignment problem do we get the luxury of arguing about what goals to select. Now let's indulge in this luxury.

Since ancient times, philosophers have dreamt of deriving ethics (principles that govern how we should behave) from scratch, using only incontrovertible principles and logic. Alas, thousands of years later, the only consensus that has been reached is that there's no consensus. For example, while Aristotle emphasized virtues, Immanuel Kant emphasized duties and utilitarians emphasized the greatest happiness for the greatest number. Kant argued that he could derive from first principles (which he called "categorical imperatives") conclusions that many contemporary philosophers disagree with: that masturbation is worse than suicide, that homosexuality is abhorrent, that it's OK to kill bastards, and that wives, servants and children are owned in a way similar to objects.

On the other hand, despite this discord, there are many ethical themes about which there's widespread agreement, both across cultures and across centuries. For example, emphasis on *beauty*, *good-*

ness and *truth* traces back to both the Bhagavad Gita and Plato. The Institute for Advanced Study in Princeton, where I once worked as a postdoc, has the motto "Truth & Beauty," while Harvard University skipped the aesthetic emphasis and went with simply "Veritas," truth. In his book *A Beautiful Question*, my colleague Frank Wilczek argues that truth is linked to beauty and that we can view our Universe as a work of art. Science, religion and philosophy all aspire to truth. Religions place strong emphasis on goodness, and so does my own university, MIT: in his 2015 commencement speech, our president, Rafael Reif, emphasized our mission to make our world a better place.

Although attempts to derive a consensus ethics from scratch have thus far failed, there's broad agreement that some ethical principles follow from more fundamental ones, as subgoals of more fundamental goals. For example, the aspiration to truth can be viewed as the quest for a better world model from figure 7.2: understanding the ultimate nature of reality helps with other ethical goals. Indeed, we now have an excellent framework for our truth quest: the scientific method. But how can we determine what's beautiful or good? Some aspects of beauty can also be traced back to underlying goals. For example, our standards of male and female beauty may partly reflect our subconscious assessment of suitability for replicating our genes.

As regards goodness, the so-called Golden Rule (that one should treat others as one would like others to treat oneself) appears in most cultures and religions, and is clearly intended to promote the harmonious continuation of human society (and hence our genes) by fostering collaboration and discouraging unproductive strife.[7] The same can be said for many of the more specific ethical rules that have been enshrined in legal systems around the world, such as the Confucian emphasis on honesty, and many of the Ten Commandments, including "Thou shalt not kill." In other words, many ethical principles have commonalities with social emotions such as empathy and compassion: they evolved to engender collaboration, and they affect our behavior through rewards and punishments. If we do something mean and feel bad about it afterward, our emotional punishment is meted out directly by our brain chemistry. If we violate ethical principles, on the other hand, society may punish us in more indirect ways such as through informal shaming by our peers or by penalizing us for breaking a law.

In other words, although humanity today is nowhere near an ethical consensus, there are many basic principles around which there's broad agreement. This agreement isn't surprising, because human societies that have survived until the present tend to have ethical principles that were optimized for the same goal: promoting their survival and flourishing. As we look ahead to a future where life has the potential to flourish throughout our cosmos for billions of years, which minimum set of ethical principles might we agree that we want this future to satisfy? This is a conversation we all need to be part of. It's been fascinating for me to hear and read the ethical views of many thinkers over many years, and the way I see it, most of their preferences can be distilled into four principles:

- Utilitarianism: Positive conscious experiences should be maximized and suffering should be minimized.
- Diversity: A diverse set of positive experiences is better than many repetitions of the same experience, even if the latter has been identified as the most positive experience possible.
- Autonomy: Conscious entities/societies should have the freedom to pursue their own goals unless this conflicts with an overriding principle.
- Legacy: Compatibility with scenarios that most humans *today* would view as happy, incompatibility with scenarios that essentially all humans *today* would view as terrible.

Let's take a moment to unpack and explore these four principles. Traditionally, utilitarianism is taken to mean "the greatest happiness for the greatest number of people," but I've generalized it here to be less anthropocentric, so that it can also include non-human animals, conscious simulated human minds, and other AIs that may exist in the future. I've made the definition in terms of *experiences* rather than people or things, because most thinkers agree that beauty, joy, pleasure and suffering are subjective experiences. This implies that if there's no experience (as in a dead universe or one populated by zombie-like unconscious machines), there can be no meaning or anything else that's ethically relevant. If we buy into this utilitarian ethical principle, then it's crucial that we figure out which intelligent systems are conscious (in the sense of having a subjective experience) and which aren't; this is the topic of the next chapter.

If this utilitarian principle was the only one we cared about, then we might wish to figure out which is the single most positive experience possible, and then settle our cosmos and re-create this exact same experience (and nothing else) over and over again, as many times as possible in as many galaxies as possible—using simulations if that's the most efficient way. If you feel that this is too banal a way to spend our cosmic endowment, then I suspect that at least part of what you find lacking in this scenario is diversity. How would you feel if all your meals for the rest of your life were identical? If all movies you ever watched were the same one? If all your friends looked identical and had identical personalities and ideas? Perhaps part of our preference for diversity stems from its having helped humanity survive and flourish, by making us more robust. Perhaps it's also linked to a preference for intelligence: the growth of intelligence during our 13.8 billion years of cosmic history has transformed boring uniformity into ever more diverse, differentiated and complex structures that process information in ever more elaborate ways.

The autonomy principle underlies many of the freedoms and rights spelled out in the Universal Declaration of Human Rights adopted by the United Nations in 1948 in an attempt to learn lessons from two world wars. This includes freedom of thought, speech and movement, freedom from slavery and torture, the right to life, liberty, security and education and the right to marry, work and own property. If we wish to be less anthropocentric, we can generalize this to the freedom to think, learn, communicate, own property and not be harmed, and the right to do whatever doesn't infringe on the freedoms of others. The autonomy principle helps with diversity, as long as everyone doesn't share exactly the same goals. Moreover, this autonomy principle follows from the utility principle if individual entities have positive experiences as goals and try to act in their own best interest: if we were instead to ban an entity from pursuing its goal even though this would cause no harm to anyone else, there would be fewer positive experiences overall. Indeed, this argument for autonomy is precisely the argument that economists use for a free market: it naturally leads to an efficient situation (called "Pareto-optimality" by economists) where nobody can get better off without someone else getting worse off.

The legacy principle basically says that we should have some say about the future since we're helping create it. The autonomy and legacy principles both embody democratic ideals: the former gives future life forms power over how the cosmic endowment gets used, while the latter gives even today's humans some power over this.

Although these four principles may sound rather uncontroversial, implementing them in practice is tricky because the devil is in the details. The trouble is reminiscent of the problems with the famous "Three Laws of Robotics" devised by sci-fi legend Isaac Asimov:

1. A robot may not injure a human being or, through inaction, allow a human being to come to harm.
2. A robot must obey the orders given it by human beings except where such orders would conflict with the First Law.
3. A robot must protect its own existence as long as such protection doesn't conflict with the First or Second Laws.

Although this all sounds good, many of Asimov's stories show how the laws lead to problematic contradictions in unexpected situations. Now suppose that we replace these laws by merely two, in an attempt to codify the autonomy principle for future life forms:

1. A conscious entity has the freedom to think, learn, communicate, own property and not be harmed or destroyed.
2. A conscious entity has the right to do whatever doesn't conflict with the first law.

Sounds good, no? But please ponder this for a moment. If animals are conscious, then what are predators supposed to eat? Must all your friends become vegetarians? If some sophisticated future computer programs turn out to be conscious, should it be illegal to terminate them? If there are rules against terminating digital life forms, then need there also be restrictions on creating them to avoid a digital population explosion? There was widespread agreement on the Universal Declaration of Human Rights simply because only humans were asked. As soon as we consider a wider range of conscious entities with varying degrees of capability and power, we face tricky trade-offs between protecting the weak and "might makes right."

There are thorny problems with the legacy principle as well. Given how ethical views have evolved since the Middle Ages regarding slavery, women's rights, etc., would we really want people from 1,500 years ago to have a lot of influence over how today's world is run? If not, why should we try to impose our ethics on future beings that may be dramatically smarter than us? Are we really confident that superhuman AGI would want what our inferior intellects cherish? This would be like a four-year-old imagining that once she grows up and gets much smarter, she's going to want to build a gigantic gingerbread house where she can spend all day eating candy and ice cream. Like her, life on Earth is likely to outgrow its childhood interests. Or imagine a mouse creating human-level AGI, and figuring it will want to build entire cities out of cheese. On the other hand, if we knew that superhuman AI would one day commit cosmocide and extinguish all life in our Universe, why should today's humans agree to this lifeless future if we have the power to prevent it by creating tomorrow's AI differently?

In conclusion, it's tricky to fully codify even widely accepted ethical principles into a form applicable to future AI, and this problem deserves serious discussion and research as AI keeps progressing. In the meantime, however, let's not let perfect be the enemy of good: there are many examples of uncontroversial "kindergarten ethics" that can and should be built into tomorrow's technology. For example, large civilian passenger aircraft shouldn't be allowed to fly into stationary objects, and now that virtually all of them have autopilot, radar and GPS, there are no longer any valid technical excuses. Yet the September 11 hijackers flew three planes into buildings and suicidal pilot Andreas Lubitz flew Germanwings Flight 9525 into a mountain on March 24, 2015—by setting the autopilot to an altitude of 100 feet (30 meters) above sea level and letting the flight computer do the rest of the work. Now that our machines are getting smart enough to have some information about what they're doing, it's time for us to teach them limits. Any engineer designing a machine needs to ask if there are things that it can but shouldn't do, and consider whether there's a practical way of making it impossible for a malicious or clumsy user to cause harm.

Ultimate Goals?

This chapter has been a brief history of goals. If we could watch a fast-forward replay of our 13.8-billion-year cosmic history, we'd witness several distinct stages of goal-oriented behavior:

1. Matter seemingly intent on maximizing its *dissipation*
2. Primitive life seemingly trying to maximize its *replication*
3. Humans pursuing not replication but goals related to pleasure, curiosity, compassion and other feelings that they'd evolved to help them replicate
4. Machines built to help humans pursue their human goals

If these machines eventually trigger an intelligence explosion, then how will this history of goals ultimately end? Might there be a goal system or ethical framework that almost all entities converge to as they get ever more intelligent? In other words, do we have an ethical destiny of sorts?

A cursory reading of human history might suggest hints of such a convergence: in his book *The Better Angels of Our Nature*, Steven Pinker argues that humanity has been getting less violent and more cooperative for thousands of years, and that many parts of the world have seen increasing acceptance of diversity, autonomy and democracy. Another hint of convergence is that the pursuit of truth through the scientific method has gained in popularity over past millennia. However, it may be that these trends show convergence not of ultimate goals but merely of subgoals. For example, figure 7.2 shows that the pursuit of truth (a more accurate world model) is simply a subgoal of almost any ultimate goal. Similarly, we saw above how ethical principles such as cooperation, diversity and autonomy can be viewed as subgoals, in that they help societies function efficiently and thereby help them survive and accomplish any more fundamental goals that they may have. Some may even dismiss everything we call "human values" as nothing but a cooperation protocol, helping us with the subgoal of collaborating more efficiently. In the same spirit, looking ahead, it's likely that any superintelligent AIs will have subgoals including efficient hardware, efficient software, truth-seeking and curiosity, simply because these subgoals help them accomplish whatever their ultimate goals are.

Indeed, Nick Bostrom argues strongly against the ethical destiny hypothesis in his book *Superintelligence*, presenting a counterpoint that he terms the *orthogonality thesis:* that the ultimate goals of a system can be independent of its intelligence. By definition, intelligence is simply the ability to accomplish complex goals, regardless of what these goals are, so the orthogonality thesis sounds quite reasonable. After all, people can be intelligent and kind or intelligent and cruel, and intelligence can be used for the goal of making scientific discoveries, creating beautiful art, helping people or planning terrorist attacks.[8]

The orthogonality thesis is empowering by telling us that the ultimate goals of life in our cosmos aren't predestined, but that we have the freedom and power to shape them. It suggests that guaranteed convergence to a unique goal is to be found not in the future but in the past, when all life emerged with the single goal of replication. As cosmic time passes, ever more intelligent minds get the opportunity to rebel and break free from this banal replication goal and choose goals of their own. We humans aren't fully free in this sense, since many goals remain genetically hardwired into us, but AIs can enjoy this ultimate freedom of being fully unfettered from prior goals. This possibility of greater goal freedom is evident in today's narrow and limited AI systems: as I mentioned earlier, the only goal of a chess computer is to win at chess, but there are also computers whose goal is to lose at chess and which compete in reverse chess tournaments where the goal is to force the opponent to capture your pieces. Perhaps this freedom from evolutionary biases can make AIs more ethical than humans in some deep sense: moral philosophers such as Peter Singer have argued that most humans behave unethically for evolutionary reasons, for example by discriminating against nonhuman animals.

We saw that a cornerstone in the "friendly-AI" vision is the idea that a recursively self-improving AI will wish to retain its ultimate (friendly) goal as it gets more intelligent. But how can an "ultimate goal" (or "final goal," as Bostrom calls it) even be defined for a superintelligence? The way I see it, we can't have confidence in the friendly-AI vision unless we can answer this crucial question.

In AI research, intelligent machines typically have a clear-cut and well-defined final goal, for instance to win the chess game or drive the car to the destination legally. The same holds for most tasks that we assign to humans, because the time horizon and context are known and limited. But now we're talking about the entire future of life in our Universe, limited by nothing but the (still not fully known) laws of physics, so defining a goal is daunting! Quantum effects aside, a truly well-defined goal would specify how all particles in our Universe should be arranged at the end of time. But it's not clear that there exists a well-defined end of time in physics. If the particles are arranged in that way at an earlier time, that arrangement will typically not last. And what particle arrangement is preferable, anyway?

We humans tend to prefer some particle arrangements over others; for example, we prefer our hometown arranged as it is over having its particles rearranged by a hydrogen bomb explosion. So suppose we try to define a *goodness function* that associates a number with every possible arrangement of the particles in our Universe, quantifying how "good" we think this arrangement is, and then give a superintelligent AI the goal of maximizing this function. This may sound like a reasonable approach, since describing goal-oriented behavior as function maximization is popular in other areas of science: for example, economists often model people as trying to maximize what they call a "utility function," and many AI designers train their intelligent agents to maximize what they call a "reward function." When we're taking about the ultimate goals for our cosmos, however, this approach poses a computational nightmare, since it would need to define a goodness value for every one of more than a googolplex possible arrangements of the elementary particles in our Universe, where a googolplex is 1 followed by 10^{100} zeroes—more zeroes than there are particles in our Universe. How would we define this goodness function to the AI?

As we've explored above, the only reason that we humans have any preferences at all may be that we're the solution to an evolutionary optimization problem. Thus all normative words in our human language, such as "delicious," "fragrant," "beautiful," "comfortable," "interesting," "sexy," "meaningful," "happy" and "good," trace their origin to this evolutionary optimization: there is therefore no guarantee that a superintelligent AI would find them rigorously definable. Even if the AI learned to accurately predict the preferences of some representative human, it wouldn't be able to compute the goodness function for most particle arrangements: the vast majority of possible particle arrangements correspond to strange cosmic scenarios with no stars, planets or people whatsoever, with which humans have no experience, so who is to say how "good" they are?

There are of course *some* functions of the cosmic particle arrangement that can be rigorously defined, and we even know of physical systems that evolve to maximize some of them. For example, we've already discussed how many systems evolve to maximize their *entropy*, which in the absence of gravity eventually leads to heat death, where everything is boringly uniform and unchanging. So entropy is hardly something we would want our AI to call "goodness" and strive to maximize. Here are a few examples of other quantities that one could strive to maximize and that may be rigorously definable in terms of particle arrangements:

- The fraction of all the matter in our Universe that's in the form of a particular organism, say humans or *E. coli* (inspired by evolutionary inclusive-fitness maximization)
- The ability of an AI to predict the future, which AI researcher Marcus Hutter argues is a good measure of its intelligence
- What AI researchers Alex Wissner-Gross and Cameron Freer term *causal entropy* (a proxy for future opportunities), which they argue is the hallmark of intelligence
- The computational capacity of our Universe

- The algorithmic complexity of our Universe (how many bits are needed to describe it)
- The amount of consciousness in our Universe (see next chapter)

However, when one starts with a physics perspective, where our cosmos consists of elementary particles in motion, it's hard to see how one rather than another interpretation of "goodness" would naturally stand out as special. We have yet to identify any final goal for our Universe that appears both definable and desirable. The only currently programmable goals that are guaranteed to remain truly well-defined as an AI gets progressively more intelligent are goals expressed in terms of physical quantities alone, such as particle arrangements, energy and entropy. However, we currently have no reason to believe that any such definable goals will be desirable in guaranteeing the survival of humanity.

Contrariwise, it appears that we humans are a historical accident, and aren't the optimal solution to any well-defined physics problem. This suggests that a superintelligent AI with a rigorously defined goal will be able to improve its goal attainment by eliminating us. This means that to wisely decide what to do about AI development, we humans need to confront not only traditional computational challenges, but also some of the most obdurate questions in philosophy. To program a self-driving car, we need to solve the trolley problem of whom to hit during an accident. To program a friendly AI, we need to capture the meaning of life. What's "meaning"? What's "life"? What's the ultimate ethical imperative? In other words, how should we strive to shape the future of our Universe? If we cede control to a superintelligence before answering these questions rigorously, the answer it comes up with is unlikely to involve us. This makes it timely to rekindle the classic debates of philosophy and ethics, and adds a new urgency to the conversation!

THE BOTTOM LINE:

- The ultimate origin of goal-oriented behavior lies in the laws of physics, which involve optimization.
- Thermodynamics has the built-in goal of *dissipation:* to increase a measure of messiness that's called *entropy*.
- *Life* is a phenomenon that can help dissipate (increase overall messiness) even faster by retaining or growing its complexity and replicating while increasing the messiness of its environment.
- Darwinian evolution shifts the goal-oriented behavior from dissipation to replication.
- Intelligence is the ability to accomplish complex goals.
- Since we humans don't always have the resources to figure out the truly optimal replication strategy, we've evolved useful rules of thumb that guide our decisions: feelings such as hunger, thirst, pain, lust and compassion.
- We therefore no longer have a simple goal such as replication; when our feelings conflict with the goal of our genes, we obey our feelings, as by using birth control.
- We're building increasingly intelligent machines to help us accomplish our goals. Insofar as we build such machines to exhibit goal-oriented behavior, we strive to align the machine goals with ours.
- Aligning machine goals with our own involves three unsolved problems: making machines learn them, adopt them and retain them.
- AI can be created to have virtually any goal, but almost any sufficiently ambitious goal can lead to subgoals of self-preservation, resource acquisition and curiosity to understand the world better—the former two may potentially lead a superintelligent AI to cause problems for humans, and the latter may prevent it from retaining the goals we give it.
- Although many broad ethical principles are agreed upon by most humans, it's unclear how to apply them to other entities, such as non-human animals and future AIs.
- It's unclear how to imbue a superintelligent AI with an ultimate goal that neither is undefined nor leads to the elimination of humanity, making it timely to rekindle research on some of the thorniest issues in philosophy!

Chapter 8

Consciousness

I cannot imagine a consistent theory of everything that ignores consciousness.

Andrei Linde, 2002

We should strive to grow consciousness itself—to generate bigger, brighter lights in an otherwise dark universe.

Giulio Tononi, 2012

We've seen that AI can help us create a wonderful future if we manage to find answers to some of the oldest and toughest problems in philosophy—by the time we need them. We face, in Nick Bostrom's words, philosophy with a deadline. In this chapter, let's explore one of the thorniest philosophical topics of all: consciousness.

Who Cares?

Consciousness is controversial. If you mention the "C-word" to an AI researcher, neuroscientist or psychologist, they may roll their eyes. If they're your mentor, they might instead take pity on you and try to talk you out of wasting your time on what they consider a hopeless and unscientific problem. Indeed, my friend Christof Koch, a renowned neuroscientist who leads the Allen Institute for Brain Science, told me that he was once warned of working on consciousness before he had tenure—by none less than Nobel laureate Francis Crick. If you look up "consciousness" in the 1989 *Macmillan Dictionary of Psychol-*

ogy, you're informed that "Nothing worth reading has been written on it."[1] As I'll explain in this chapter, I'm more optimistic!

Although thinkers have pondered the mystery of consciousness for thousands of years, the rise of AI adds a sudden urgency, in particular to the question of predicting which intelligent entities have subjective experiences. As we saw in chapter 3, the question of whether intelligent machines should be granted some form of rights depends crucially on whether they're conscious and can suffer or feel joy. As we discussed in chapter 7, it becomes hopeless to formulate utilitarian ethics based on maximizing positive experiences without knowing which intelligent entities are capable of having them. As mentioned in chapter 5, some people might prefer their robots to be unconscious to avoid feeling slave-owner guilt. On the other hand, they may desire the opposite if they upload their minds to break free from biological limitations: after all, what's the point of uploading yourself into a robot that talks and acts like you if it's a mere unconscious zombie, by which I mean that being the uploaded you doesn't feel like anything? Isn't this equivalent to committing suicide from your subjective point of view, even though your friends may not realize that your subjective experience has died?

For the long-term cosmic future of life (chapter 6), understanding what's conscious and what's not becomes pivotal: if technology enables intelligent life to flourish throughout our Universe for billions of years, how can we be sure that this life is conscious and able to appreciate what's happening? If not, then would it be, in the words of the famous physicist Erwin Schrödinger, "a play before empty benches, not existing for anybody, thus quite properly speaking not existing"?[2] In other words, if we enable high-tech descendants that we mistakenly think are conscious, would this be the ultimate zombie apocalypse, transforming our grand cosmic endowment into nothing but an astronomical waste of space?

What Is Consciousness?

Many arguments about consciousness generate more heat than light because the antagonists are talking past each other, unaware that they're using different definitions of the C-word. Just as with "life" and "intelligence," there's no undisputed correct definition of the word "consciousness." Instead, there are many competing ones, including sentience, wakefulness, self-awareness, access to sensory input and ability to fuse information into a narrative.[3] In our exploration of the future of intelligence, we want to take a maximally broad and inclusive view, not limited to the sorts of biological consciousness that exist so far. That's why the definition I gave in chapter 1, which I'm sticking with throughout this book, is very broad:

consciousness = *subjective experience*

In other words, if it feels like something to be you right now, then you're conscious. It's this particular definition of consciousness that gets to the crux of all the AI-motivated questions in the previous section: Does it feel like something to be Prometheus, AlphaGo or a self-driving Tesla?

To appreciate how broad our consciousness definition is, note that it doesn't mention behavior, perception, self-awareness, emotions or attention. So by this definition, you're conscious also when you're dreaming, even though you lack wakefulness or access to sensory input and (hopefully!) aren't sleepwalking and doing things. Similarly, any system that experiences pain is conscious in this sense, even if it can't move. Our definition leaves open the possibility that some future AI systems may be conscious too, even if they exist merely as software and aren't connected to sensors or robotic bodies.

With this definition, it's hard not to care about consciousness. As Yuval Noah Harari puts it in his book *Homo Deus*:[4] "If any scientist wants to argue that subjective experiences are irrelevant, their challenge is to explain why torture or rape are wrong without reference to any subjective experience." Without such reference, it's all just a bunch of elementary particles moving around according to the laws of physics—and what's wrong with that?

What's the Problem?

So what precisely is it that we don't understand about consciousness? Few have thought harder about this question than David Chalmers, a famous Australian philosopher rarely seen without a playful smile and a black leather jacket—which my wife liked so much that she gave me a similar one for Christmas. He followed his heart into philosophy despite making the finals at the International Mathematics Olympiad—and despite the fact that his only B grade in college, shattering his otherwise straight As, was for an introductory philosophy course. Indeed, he seems utterly undeterred by put-downs or controversy, and I've been astonished by his ability to politely listen to uninformed and misguided criticism of his own work without even feeling the need to respond.

As David has emphasized, there are really two separate mysteries of the mind. First, there's the mystery of how a brain processes information, which David calls the "easy" problems. For example, how does a brain attend to, interpret and respond to sensory input? How can it report on its internal state using language? Although these questions are actually extremely difficult, they're by our definitions not mysteries of consciousness, but mysteries of intelligence: they ask how a brain remembers, computes and learns. Moreover, we saw in the first part of the book how AI researchers have started to make serious progress on solving many of these "easy problems" with machines—from playing Go to driving cars, analyzing images and processing natural language.

Then there's the separate mystery of why you have a subjective experience, which David calls the *hard* problem. When you're driving, you're experiencing colors, sounds, emotions, and a feeling of self. But why are you experiencing anything at all? Does a self-driving car experience anything at all? If you're racing against a self-driving car, you're both inputting information from sensors, processing it and outputting motor commands. But subjectively *experiencing* driving is something logically separate—is it optional, and if so, what causes it?

I approach this hard problem of consciousness from a physics point of view. From my perspective, a conscious person is simply food, rearranged. So why is one arrangement conscious, but not the other?

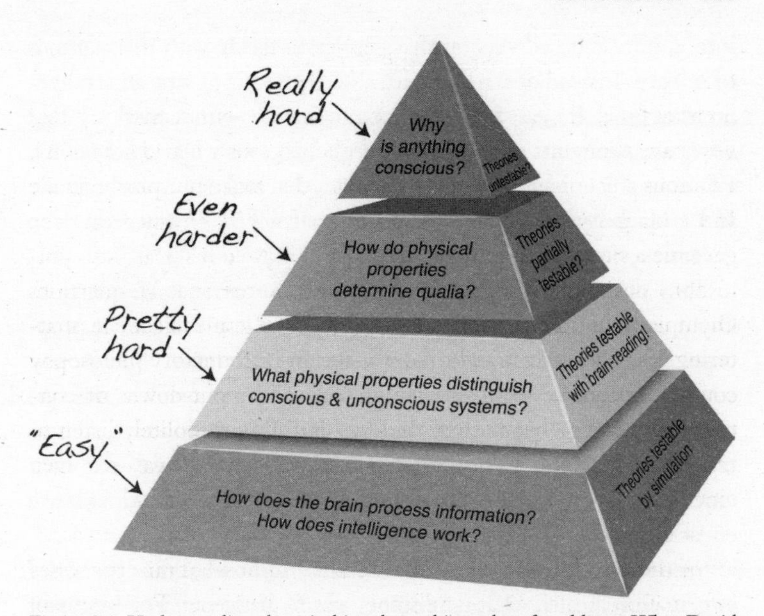

Figure 8.1: Understanding the mind involves a hierarchy of problems. What David Chalmers calls the "easy" problems can be posed without mentioning subjective experience. The apparent fact that some but not all physical systems are conscious poses three separate questions. If we have a theory for answering the question that defines the "pretty hard problem," then it can be experimentally tested. If it works, then we can build on it to tackle the tougher questions above.

Moreover, physics teaches us that food is simply a large number of quarks and electrons, arranged in a certain way. So which particle arrangements are conscious and which aren't?*

What I like about this physics perspective is that it transforms the hard problem that we as humans have struggled with for millennia

* An alternative viewpoint is *substance dualism*—that living entities differ from inanimate ones because they contain some non-physical substance such as an "anima," "élan vital" or "soul." Support for substance dualism among scientists has gradually dwindled. To understand why, consider that your body is made of about 10^{29} quarks and electrons, which, as far as we can tell, move according to simple physical laws. Imagine a future technology able to track all your particles: if they were found to obey the laws of physics exactly, then your purported soul is having no effect on your particles, so your conscious mind and its ability to control your movements would have nothing to do with a soul. If your particles were instead found not to obey the known laws of physics because they were being pushed around by your soul, then the new entity causing these forces would by definition be a physical one that we can study just like we've studied new fields and new particles in the past.

into a more focused version that's easier to tackle with the methods of science. Instead of starting with a hard *problem* of why an arrangement of particles can feel conscious, let's start with a hard *fact* that some arrangements of particles do feel conscious while others don't. For example, you know that the particles that make up your brain are in a conscious arrangement right now, but not when you're in deep dreamless sleep.

This physics perspective leads to three separate hard questions about consciousness, as shown in figure 8.1. First of all, what properties of the particle arrangement make the difference? Specifically, what physical properties distinguish conscious and unconscious systems? If we can answer that, then we can figure out which AI systems are conscious. In the more immediate future, it can also help emergency-room doctors determine which unresponsive patients are conscious.

Second, how do physical properties determine what the experience is like? Specifically, what determines *qualia*, basic building blocks of consciousness such as the redness of a rose, the sound of a cymbal, the smell of a steak, the taste of a tangerine or the pain of a pinprick?*

Third, why is anything conscious? In other words, is there some deep undiscovered explanation for why clumps of matter can be conscious, or is this just an unexplainable brute fact about the way the world works?

The computer scientist Scott Aaronson, a former MIT colleague of mine, has lightheartedly called the first question the "pretty hard problem" (PHP), as has David Chalmers. In that spirit, let's call the other two the "even harder problem" (EHP) and the "really hard problem" (RHP), as illustrated in figure 8.1.†

* I use the word "qualia" according to the dictionary definition, to mean individual instances of subjective experience—that is, to mean the subjective experience itself, not any purported substance causing the experience. Beware that some people use the word differently.

† I'd originally called the RHP the "very hard problem," but after I showed this chapter to David Chalmers, he emailed me the clever suggestion of switching to the "*really hard problem,*" to match what he really meant: "since the first two problems (at least put this way) aren't really part of the hard problem as I conceived of it whereas the third problem is, you could perhaps use 'really hard' instead of 'very hard' for the third one to match my usage."

Is Consciousness Beyond Science?

When people tell me that consciousness research is a hopeless waste of time, the main argument they give is that it's "unscientific" and always will be. But is that really true? The influential Austro-British philosopher Karl Popper popularized the now widely accepted adage "If it's not falsifiable, it's not scientific." In other words, science is all about testing theories against observations: if a theory can't be tested even in principle, then it's logically impossible to ever falsify it, which by Popper's definition means that it's unscientific.

So could there be a scientific theory that answers any of the three consciousness questions in figure 8.1? Please let me try to persuade you that the answer is a resounding YES!, at least for the pretty hard problem: "What physical properties distinguish conscious and unconscious systems?" Suppose that someone has a theory that, given any physical system, answers the question of whether the system is conscious with "yes," "no" or "unsure." Let's hook your brain up to a device that measures some of the information processing in different parts of your brain, and let's feed this information into a computer program that uses the consciousness theory to predict which parts of that information are conscious, and presents you with its predictions in real time on a screen, as in figure 8.2. First you think of an apple. The screen informs you that there's information about an apple in your brain which you're aware of, but that there's also information in your brainstem about your pulse that you're unaware of. Would you be impressed? Although the first two predictions of the theory were correct, you decide to do some more rigorous testing. You think about your mother and the computer informs you that there's information in your brain about your mother but that you're unaware of this. The theory made an incorrect prediction, which means that it's ruled out and goes in the garbage dump of scientific history together with Aristotelian mechanics, the luminiferous aether, geocentric cosmology and countless other failed ideas. Here's the key point: Although the theory was wrong, it was *scientific*! Had it not been scientific, you wouldn't have been able to test it and rule it out.

Someone might criticize this conclusion and say that *they* have no evidence of what you're conscious of, or even of you being conscious

Figure 8.2: Suppose that a computer measures information being processed in your brain and predicts which parts of it you're aware of according to a theory of consciousness. You can scientifically test this theory by checking whether its predictions are correct, matching your subjective experience.

at all: although they heard you say that you're conscious, an unconscious zombie could conceivably say the same thing. But this doesn't make that consciousness theory unscientific, because they can trade places with you and test whether it correctly predicts *their own* conscious experiences.

On the other hand, if the theory refuses to make any predictions, merely replying "unsure" whenever queried, then it's untestable and hence unscientific. This might happen because it's applicable only in some situations, because the required computations are too hard to carry out in practice or because the brain sensors are no good. Today's most popular scientific theories tend to be somewhere in the middle, giving testable answers to some but not all of our questions. For example, our core theory of physics will refuse to answer questions about systems that are simultaneously extremely small (requiring quantum mechanics) and extremely heavy (requiring general relativity), because we haven't yet figured out which mathematical equations to use in this case. This core theory will also refuse to predict the exact masses of all possible atoms—in this case, we think we have the necessary equations, but we haven't managed to accurately compute their solutions. The more dangerously a theory lives by sticking its neck out and making testable predictions, the more useful it is, and the more seriously we take it if it survives all our attempts to kill it. Yes, we can only test *some* predictions of consciousness theories, but that's how it is for *all* physical theories. So let's not waste time whining about what we can't test, but get to work testing what we *can* test!

In summary, any theory predicting which physical systems are conscious (the pretty hard problem) is scientific, as long as it can predict which of your brain processes are conscious. However, the testability issue becomes less clear for the higher-up questions in figure 8.1. What would it mean for a theory to predict how you subjectively experience the color red? And if a theory purports to explain why there is such a thing as consciousness in the first place, then how do you test it experimentally? Just because these questions are hard doesn't mean that we should avoid them, and we'll indeed return to them below. But when confronted with several related unanswered questions, I think it's wise to tackle the easiest one first. For this reason, my consciousness research at MIT is focused squarely on the base of the pyramid in figure 8.1. I recently discussed this strategy with my fellow physicist Piet Hut from Princeton, who joked that trying to build the top of the pyramid before the base would be like worrying about the interpretation of quantum mechanics before discovering the Schrödinger equation, the mathematical foundation that lets us predict the outcomes of our experiments.

When discussing what's beyond science, it's important to remember that the answer depends on time! Four centuries ago, Galileo Galilei was so impressed by math-based physics theories that he described nature as "a book written in the language of mathematics." If he threw a grape and a hazelnut, he could accurately predict the shapes of their trajectories and when they would hit the ground. Yet he had no clue why one was green and the other brown, or why one was soft and the other hard—these aspects of the world were beyond the reach of science at the time. But not forever! When James Clerk Maxwell discovered his eponymous equations in 1861, it became clear that light and colors could also be understood mathematically. We now know that the aforementioned Schrödinger equation, discovered in 1925, can be used to predict all properties of matter, including what's soft or hard. While theoretical progress has enabled ever more scientific predictions, technological progress has enabled ever more experimental tests: almost everything we now study with telescopes, microscopes or particle colliders was once beyond science. In other words, the purview of science has expanded dramatically since Galileo's days, from a tiny fraction of all phenomena to a large percentage, including subatomic particles, black holes and our cosmic origins 13.8 billion years ago. This raises the question: What's left?

To me, consciousness is the elephant in the room. Not only do you know that you're conscious, but it's *all* you know with complete certainty—everything else is inference, as René Descartes pointed out back in Galileo's time. Will theoretical and technological progress eventually bring even consciousness firmly into the domain of science? We don't know, just as Galileo didn't know whether we'd one day understand light and matter.* Only one thing is guaranteed: we won't succeed if we don't try! That's why I and many other scientists around the world are trying hard to formulate and test theories of consciousness.

Experimental Clues About Consciousness

Lots of information processing is taking place in our heads right now. Which of it is conscious and which isn't? Before exploring consciousness theories and what they predict, let's look at what experiments have taught us so far, ranging from traditional low-tech or no-tech observations to state-of-the-art brain measurements.

What Behaviors Are Conscious?

If you multiply 32 by 17 in your head, you're conscious of many of the inner workings of your computation. But suppose I instead show you a portrait of Albert Einstein and tell you to say the name

* If our physical reality is entirely mathematical (information-based, loosely speaking), as I explored in my book *Our Mathematical Universe*, then no aspect of reality—not even consciousness—lies beyond the purview of science. Indeed, the really hard problem of consciousness is, from that perspective, the exact same problem as that of understanding how something mathematical can feel physical: if part of a mathematical structure is conscious, then it will experience the rest as its external physical world.

of its subject. As we saw in chapter 2, this too is a computational task: your brain is evaluating a function whose input is information from your eyes about a large number of pixel colors and whose output is information to muscles controlling your mouth and vocal cords. Computer scientists call this task "image classification" followed by "speech synthesis." Although this computation is way more complicated than your multiplication task, you can do it much faster, seemingly without effort, and without being conscious of the details of *how* you do it. Your subjective experience consists merely of looking at the picture, experiencing a feeling of recognition and hearing yourself say "Einstein."

Psychologists have long known that you can unconsciously perform a wide range of other tasks and behaviors as well, from blink reflexes to breathing, reaching, grabbing and keeping your balance. Typically, you're consciously aware of what you did, but not how you did it. On the other hand, behaviors that involve unfamiliar situations, self-control, complicated logical rules, abstract reasoning or manipulation of language tend to be conscious. They're known as *behavioral correlates of consciousness*, and they're closely linked to the effortful, slow and controlled way of thinking that psychologists call "System 2."[5]

It's also known that you can convert many routines from conscious to unconscious through extensive practice, for example walking, swimming, bicycling, driving, typing, shaving, shoe tying, computer-gaming and piano playing.[6] Indeed, it's well known that experts do their specialties best when they're in a state of "flow," aware only of what's happening at a higher level, and unconscious of the low-level details of how they're doing it. For example, try reading the next sentence while being consciously aware of every single letter, as when you first learned to read. Can you feel how much slower it is, compared to when you're merely conscious of the text at the level of words or ideas?

Indeed, unconscious information processing appears not only to be possible, but also to be more the rule than the exception. Evidence suggests that of the roughly 10^7 bits of information that enter our brain each second from our sensory organs, we can be aware only of a tiny fraction, with estimates ranging from 10 to 50 bits.[7] This suggests that the information processing that we're consciously aware of is merely the tip of the iceberg.

Taken together, these clues have led some researchers to suggest that conscious information processing should be thought of as the CEO of our mind, dealing with only the most important decisions requiring complex analysis of data from all over the brain.[8] This would explain why, just like the CEO of a company, it usually doesn't want to be distracted by knowing everything its underlings are up to—but it can find them out if desired. To experience this selective attention in action, look at that word "desired" again: fix your gaze on the dot over the "i" and, without moving your eyes, shift your attention from the dot to the whole letter and then to the whole word. Although the information from your retina stayed the same, your conscious experience changed. The CEO metaphor also explains why expertise becomes unconscious: after painstakingly figuring out how to read and type, the CEO delegates these routine tasks to unconscious subordinates to be able to focus on new higher-level challenges.

Where Is Consciousness?

Clever experiments and analyses have suggested that consciousness is limited not merely to certain behaviors, but also to certain parts of the brain. Which are the prime suspects? Many of the first clues came from patients with brain lesions: localized brain damage caused by accidents, strokes, tumors or infections. But this was often inconclusive. For example, does the fact that lesions in the back of the brain can cause blindness mean that this is the site of visual consciousness, or does it merely mean that visual information passes through there en route to wherever it will later become conscious, just as it first passes through the eyes?

Although lesions and medical interventions haven't pinpointed the locations of conscious experiences, they've helped narrow down the options. For example, I know that although I experience pain in my hand as actually occurring there, the pain experience must occur elsewhere, because a surgeon once switched off my hand pain without doing anything to my hand: he merely anesthetized nerves in my shoulder. Moreover, some amputees experience phantom pain that feels as though it's in their nonexistent hand. As another example, I once noticed that when I looked only with my right eye, part of my visual field was missing—a doctor determined that my retina was coming loose and reattached it. In contrast, patients with certain brain lesions experience *hemineglect*, where they too miss information from half their visual field, but aren't even aware that it's missing—for example, failing to notice and eat the food on the left half of their plate. It's as if consciousness about half of their world has disappeared. But are those damaged brain areas supposed to generate the spatial experience, or were they merely feeding spatial information to the sites of consciousness, just as my retina did?

The pioneering U.S.-Canadian neurosurgeon Wilder Penfield found in the 1930s that his neurosurgery patients reported different parts of their body being touched when he electrically stimulated specific brain areas in what's now called the *somatosensory cortex* (figure 8.3).[9] He also found that they involuntarily moved different parts of their body when he stimulated brain areas in what's now called the *motor cortex*. But does that mean that information processing in these brain areas corresponds to consciousness of touch and motion?

Fortunately, modern technology is now giving us much more detailed clues. Although we're still nowhere near being able to measure every single firing of all of your roughly hundred billion neurons, brain-reading technology is advancing rapidly, involving techniques with intimidating names such as fMRI, EEG, MEG, ECoG, ePhys and fluorescent voltage sensing. fMRI, which stands for functional magnetic resonance imaging, measures the magnetic properties of hydrogen nuclei to make a 3-D map of your brain roughly every second, with millimeter resolution. EEG (electroencephalography) and MEG (magnetoencephalography) measure the electric and magnetic field outside your head to map your brain thousands of times per second, but with poor resolution, unable to distinguish features smaller than a few centimeters. If you're squeamish, you'll appreciate that these three techniques are all noninvasive. If you don't mind opening up your skull, you have additional options. ECoG (electrocorticography) involves placing say a hundred wires on the surface of your brain, while ePhys (electrophysiology) involves inserting microwires, which are sometimes thinner than a human hair, deep into the brain to record voltages from as many as a thousand simultaneous locations. Many epileptic patients spend days in the hospital while ECoG is used to figure out what part of their brain is triggering seizures and should be resected, and kindly agree to let neuroscientists perform consciousness experiments on them in the meantime. Finally, fluorescent voltage sensing involves genetically manipulating neurons to emit flashes of light when firing, enabling their activity to be measured with a microscope. Out of all the techniques, it has the potential to rapidly monitor the largest number of neurons, at least in animals with transparent brains—such as the *C. elegans* worm with its 302 neurons and the larval zebrafish with its about 100,000.

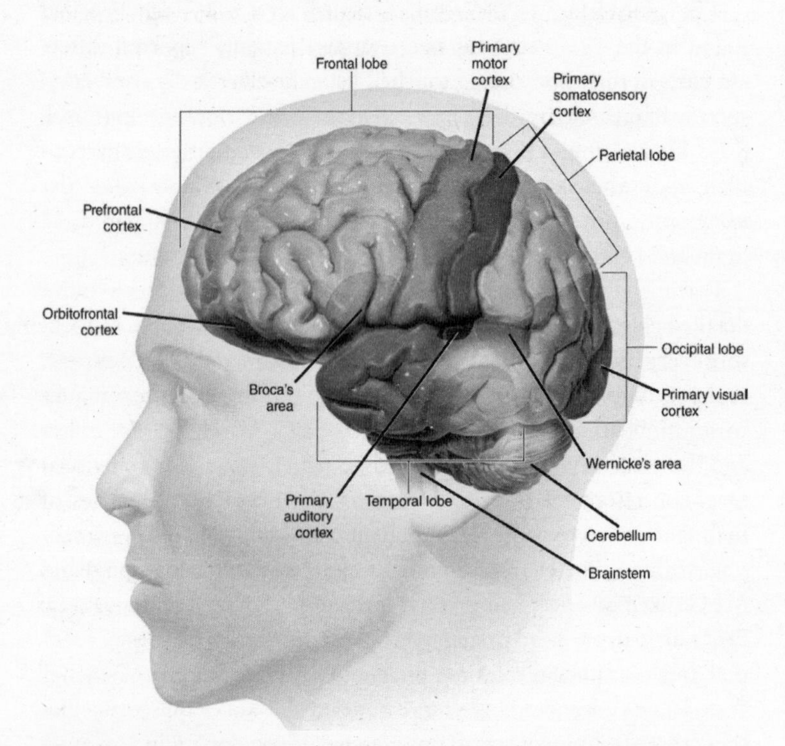

Figure 8.3: The visual, auditory, somatosensory and motor cortices are involved with vision, hearing, the sense of touch and motion activation, respectively—but that doesn't prove they're where *consciousness* of vision, hearing, touch and motion occurs. Indeed, recent research suggests that the primary visual cortex is completely unconscious, together with the cerebellum and brainstem. Image courtesy of Lachina (www.lachina.com).

Although Francis Crick warned Christof Koch about studying consciousness, Christof refused to give up and and eventually won Francis over. In 1990, they wrote a seminal paper about what they called "neural correlates of consciousness" (NCCs), asking which specific brain processes corresponded to conscious experiences. For thousands of years, thinkers had had access to the information processing in their brains only via their subjective experience and behavior. Crick and Koch pointed out that brain-reading technology was suddenly providing independent access to this information, allowing scientific study of which information processing corresponded to what conscious experience. Sure enough, technology-driven measurements have by now turned the quest for NCCs into quite a mainstream part of neuroscience, one whose thousands of publications extend into even the most prestigious journals.[10]

What are the conclusions so far? To get a flavor for NCC detective work, let's first ask whether your retina is conscious, or whether it's merely a zombie system that records visual information, processes it and sends it on to a system downstream in your brain where your subjective visual experience occurs. In the left panel of figure 8.4, which square is darker: the one labeled A or B? A, right? No, they're in fact identically colored, which you can verify by looking at them through small holes between your fingers. This proves that your visual experience can't reside entirely in your retina, since if it did, they'd look the same.

Now look at the right panel of figure 8.4. Do you see two women or a vase? If you look long enough, you'll subjectively experience both in succession, even though the information reaching your retina remains the same. By measuring what happens in your brain during the two situations, one can tease apart what makes the difference—and it's not the retina, which behaves identically in both cases.

The death blow to the conscious-retina hypothesis comes from a technique called "continuous flash suppression" pioneered by Christof Koch, Stanislas Dehaene and collaborators: it's been discovered that if you make one of your eyes watch a complicated sequence of rapidly changing patterns, then this will distract your visual system to such an extent that you'll be completely unaware of a still image shown to the other eye.[11] In summary, you can have a visual image

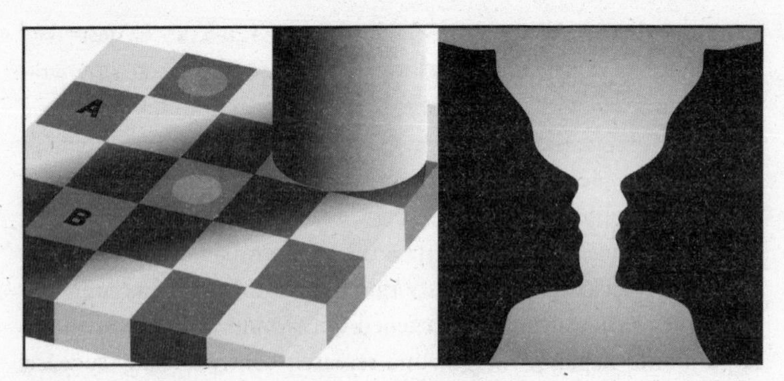

Figure 8.4: Which square is darker—A or B? What do you see on the right—a vase, two women or both in succession? Illusions such as these prove that your visual consciousness can't be in your eyes or other early stages of your visual system, because it doesn't depend only on what's in the picture.

in your retina without experiencing it, and you can (while dreaming) experience an image without it being on your retina. This proves that your two retinas don't host your visual consciousness any more than a video camera does, even though they perform complicated computations involving over a hundred million neurons.

NCC researchers also use continuous flash suppression, unstable visual/auditory illusions and other tricks to pinpoint which of your brain regions *are* responsible for each of your conscious experiences. The basic strategy is to compare what your neurons are doing in two situations where essentially everything (including your sensory input) is the same—except your conscious experience. The parts of your brain that are measured to behave differently are then identified as NCCs.

Such NCC research has proven that none of your consciousness resides in your gut, even though that's the location of your enteric nervous system with its whopping half-billion neurons that compute how to optimally digest your food; feelings such as hunger and nausea are instead produced in your brain. Similarly, none of your consciousness appears to reside in the brainstem, the bottom part of the brain that connects to the spinal cord and controls breathing, heart rate and blood pressure. More shockingly, your consciousness doesn't appear to extend to your cerebellum (figure 8.3), which contains about two-thirds of all your neurons: patients whose cerebellum is destroyed experience slurred speech and clumsy motion reminiscent of a drunkard, but remain fully conscious.

The question of which parts of your brain *are* responsible for consciousness remains open and controversial. Some recent NCC research suggests that your consciousness mainly resides in a "hot zone" involving the thalamus (near the middle of your brain) and the rear part of the cortex (the outer brain layer consisting of a crumpled-up six-layer sheet which, if flattened out, would have the area of a large dinner napkin).[12] This same research controversially suggests that the primary visual cortex at the very back of the head is an exception to this, being as unconscious as your eyeballs and your retinas.

When Is Consciousness?

So far, we've looked at experimental clues regarding what types of information processing are conscious and where consciousness occurs. But *when* does it occur? When I was a kid, I used to think that we become conscious of events as they happen, with absolutely no time lag or delay. Although that's still how it subjectively feels to me, it clearly can't be correct, since it takes time for my brain to process the information that enters via my sensory organs. NCC researchers have carefully measured how long, and Christof Koch's summary is that it takes about a quarter of a second from when light enters your eye from a complex object until you consciously perceive seeing it as what it is.[13] This means that if you're driving down a highway at fifty-five miles per hour and suddenly see a squirrel a few meters in front of you, it's too late for you to do anything about it, because you've already run over it!

In summary, your consiousness lives in the past, with Christof Koch estimating that it lags behind the outside world by about a quarter second. Intriguingly, you can often react to things faster than you can become conscious of them, which proves that the information processing in charge of your most rapid reactions must be unconscious. For example, if a foreign object approaches your eye, your blink reflex can close your eyelid within a mere tenth of a second. It's as if one of your brain systems receives ominous information from the visual system, computes that your eye is in danger of getting struck, emails your eye muscles instructions to blink and simultaneously emails the conscious part of your brain saying "Hey, we're going to blink." By the time this email has been read and included into your conscious experience, the blink has already happened.

Indeed, the system that reads that email is continually bombarded with messages from all over your body, some more delayed than others. It takes longer for nerve signals to reach your brain from your fingers than from your face because of distance, and it takes longer for you to analyze images than sounds because it's more complicated—which is why Olympic races are started with a bang rather than with a visual cue. Yet if you touch your nose, you consciously experience the sensation on your nose and fingertip as simultaneous, and if you clap your hands, you see, hear and feel the clap at exactly the same time.[14] This means that your full conscious experience of an event isn't created until the last slowpoke email reports have trickled in and been analyzed.

A famous family of NCC experiments pioneered by physiologist Benjamin Libet has shown that the sort of actions you can perform unconsciously aren't limited to rapid responses such as blinks and ping-pong smashes, but also include certain decisions that you might attribute to free will—brain measurements can sometimes predict your decision before you become conscious of having made it.[15]

Theories of Consciousness

We've just seen that, although we still don't understand consciousness, we have amazing amounts of experimental data about various aspects of it. But all this data comes from *brains*, so how can it teach us anything about consciousness in *machines*? This requires a major extrapolation beyond our current experimental domain. In other words, it requires a *theory*.

Why a Theory?

To appreciate why, let's compare theories of consciousness with theories of gravity. Scientists started taking Newton's theory of gravity seriously because they got more out of it than they put into it: simple equations that fit on a napkin could accurately predict the outcome of every gravity experiment ever conducted. They therefore also took seriously its predictions far beyond the domain where it had been tested, and these bold extrapolations turned out to work even for the motions of galaxies in clusters millions of light-years across. However, the predictions were off by a tiny amount for the motion of Mercury around the Sun. Scientists then started taking seriously Einstein's improved theory of gravity, general relativity, because it was arguably even more elegant and economical, and correctly predicted even what Newton's theory got wrong. They consequently took seriously also its predictions far beyond the domain where it had been tested, for phenomena as exotic as black holes, gravitational waves in the very fabric of spacetime, and the expansion of our Universe from a hot fiery origin—all of which were subsequently confirmed by experiment.

Analogously, if a mathematical theory of consciousness whose equations fit on a napkin could successfully predict the outcomes of all experiments we perform on brains, then we'd start taking seriously not merely the theory itself, but also its predictions for consciousness beyond brains—for example, in machines.

Consciousness from a Physics Perspective

Although some theories of consciousness date back to antiquity, most modern ones are grounded in neuropsychology and neuroscience, attempting to explain and predict consciousness in terms of neural events occurring in the brain.[16] Although these theories have made some successful predictions for neural correlates of consciousness, they neither can nor aspire to make predictions about machine consciousness. To make the leap from brains to machines, we need to generalize from NCCs to PCCs: *physical correlates of consciousness*, defined as the patterns of moving particles that are conscious. Because if a theory can correctly predict what's conscious and what's not by referring only to physical building blocks such as elementary particles and force fields, then it can make predictions not merely for brains, but also for any other arrangements of matter, including future AI systems. So let's take a physics perspective: What particle arrangements are conscious?

But this really raises another question: How can something as complex as consciousness be made of something as simple as particles? I think it's because it's a phenomenon that has properties above and beyond those of its particles. In physics, we call such phenomena "emergent."[17] Let's understand this by looking at an emergent phenomenon that's simpler than consciousness: wetness.

A drop of water is wet, but an ice crystal and a cloud of steam aren't, even though they're made of identical water molecules. Why? Because the property of wetness depends only on the arrangement of the molecules. It makes absolutely no sense to say that a single water molecule is wet, because the phenomenon of wetness emerges only when there are many molecules, arranged in the pattern we call liquid. So solids, liquids and gases are all emergent phenomena: they're more than the sum of their parts, because they have properties above and beyond the properties of their particles. They have properties that their particles lack.

Now just like solids, liquids and gases, I think consciousness is an emergent phenomenon, with properties above and beyond those of its particles. For example, entering deep sleep extinguishes consciousness, by merely rearranging the particles. In the same way, my consciousness would disappear if I froze to death, which would rearrange my particles in a more unfortunate way.

When you put lots of particles together to make anything from water to a brain, new phenomena with observable properties emerge. We physicists love studying these emergent properties, which can often be identified by a small set of numbers that you can go out and measure—quantities such as how viscous the substance is, how compressible it is and so on. For example, if a substance is so viscous that it's rigid, we call it a solid, otherwise we call it a fluid. And if a fluid isn't compressible, we call it a liquid, otherwise we call it a gas or a plasma, depending on how well it conducts electricity.

Consciousness as Information

So could there be analogous quantities that quantify consciousness? The Italian neuroscientist Giulio Tononi has proposed one such quantity, which he calls the "*integrated information*," denoted by the Greek letter Φ (*Phi*), which basically measures how much different parts of a system know about each other (see figure 8.5).

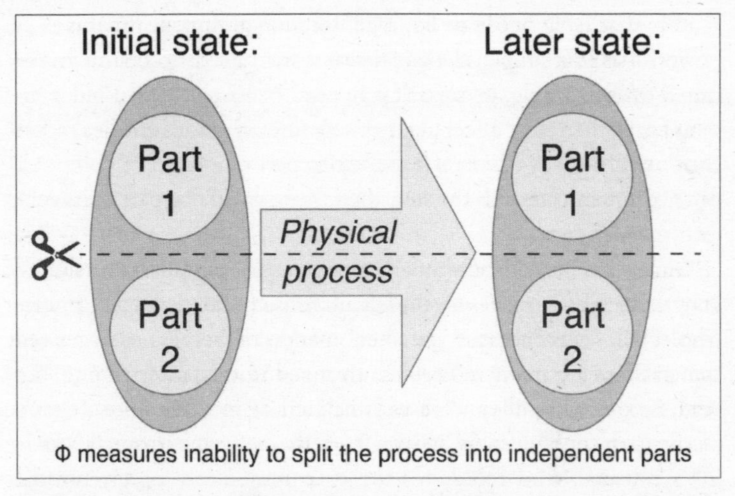

Figure 8.5: Given a physical process that, with the passage of time, transforms the initial state of a system into a new state, its *integrated information* Φ measures inability to split the process into independent parts. If the future state of each part depends only on its own past, not on what the other part has been doing, then $\Phi = 0$: what we called one system is really two independent systems that don't communicate with each other at all.

I first met Giulio at a 2014 physics conference in Puerto Rico to which I'd invited him and Christof Koch, and he struck me as the ultimate renaissance man who'd have blended right in with Galileo and Leonardo da Vinci. His quiet demeanor couldn't hide his incredible knowledge of art, literature and philosophy, and his culinary reputation preceded him: a cosmopolitan TV journalist had recently told me how Giulio had, in just a few minutes, whipped up the most delicious salad he'd tasted in his life. I soon realized that behind his soft-spoken demeanor was a fearless intellect who'd follow the evidence wherever it took him, regardless of the preconceptions and taboos of the establishment. Just as Galileo had pursued his mathematical theory of motion despite establishment pressure not to challenge geocentrism, Giulio had developed the most mathematically precise consciousness theory to date, *integrated information theory* (IIT).

I'd been arguing for decades that consciousness is the way information feels when being processed in certain complex ways.[18] IIT agrees with this and replaces my vague phrase "certain complex ways" by a precise definition: the information processing needs to be integrated, that is, Φ needs to be large. Giulio's argument for this is as powerful as it is simple: the conscious system needs to be integrated into a unified whole, because if it instead consisted of two independent parts, then they'd feel like two separate conscious entities rather than one. In other words, if a conscious part of a brain or computer can't communicate with the rest, then the rest can't be part of its subjective experience.

Giulio and his collaborators have measured a simplified version of Φ by using EEG to measure the brain's response to magnetic stimulation. Their "consciousness detector" works really well: it determined that patients were conscious when they were awake or dreaming, but unconscious when they were anesthetized or in deep sleep. It even discovered consciousness in two patients suffering from "locked-in" syndrome, who couldn't move or communicate in any normal way.[19] So this is emerging as a promising technology for doctors in the future to figure out whether certain patients are conscious or not.

Anchoring Consciousness in Physics

IIT is defined only for discrete systems that can be in a finite number of states, for example bits in a computer memory or oversimplified neurons that can be either on or off. This unfortunately means that IIT isn't defined for most traditional physical systems, which can change continuously—for example, the position of a particle or the strength of a magnetic field can take any of an infinite number of values.[20] If you try to apply the IIT formula to such systems, you'll typically get the unhelpful result that Φ is infinite. Quantum-mechanical systems can be discrete, but the original IIT isn't defined for quantum systems. So how can we anchor IIT and other information-based consciousness theories on a solid physical foundation?

We can do this by building on what we learned in chapter 2 about how clumps of matter can have emergent properties that are related to information. We saw that for something to be usable as a memory device that can store information, it needs to have many long-lived states. We also saw that being *computronium*, a substance that can do computations, in addition requires complex dynamics: the laws of physics need to make it change in ways that are complicated enough to be able to implement arbitrary information processing. Finally, we saw how a neural network, for example, is a powerful substrate for learning because, simply by obeying the laws of physics, it can rearrange itself to get better and better at implementing desired computations. Now we're asking an additional question: What makes a blob of matter able to have a subjective experience? In other words, under what conditions will a blob of matter be able to do these four things?

1. remember
2. compute
3. learn
4. experience

We explored the first three in chapter 2, and are now tackling the fourth. Just as Margolus and Toffoli coined the term *computronium* for a substance that can perform arbitrary computations, I like to use the term *sentronium* for the most general substance that has subjective experience (is sentient).*

But how can consciousness feel so non-physical if it's in fact a physical phenomenon? How can it feel so independent of its physical substrate? I think it's because it *is* rather independent of its physical substrate, the stuff in which it is a pattern! We encountered many beautiful examples of substrate-independent patterns in chapter 2, including waves, memories and computations. We saw how they weren't merely more than their parts (emergent), but rather independent of their parts, taking on a life of their own. For example, we saw how a future simulated mind or computer-game character would have no way of knowing whether it ran on Windows, Mac OS, an Android phone or some other operating system, because it would be substrate-independent. Nor could it tell whether the logic gates of its computer were made of transistors, optical circuits or other hardware. Or what the fundamental laws of physics are—they could be anything as long as they allow the construction of universal computers.

In summary, I think that consciousness is a physical phenomenon

* Although I've earlier used "perceptronium" as a synonym for sentronium, that name suggests too narrow a definition, since percepts are merely those subjective experiences that we perceive based on sensory input—excluding, for example, dreams and internally generated thoughts.

that feels non-physical because it's like waves and computations: it has properties independent of its specific physical substrate. This follows logically from the consciousness-as-information idea. This leads to a radical idea that I really like: If consciousness is the way that information feels when it's processed in certain ways, then it must be substrate-independent; it's only the structure of the information processing that matters, not the structure of the matter doing the information processing. In other words, consciousness is substrate-independent twice over!

As we've seen, physics describes patterns in spacetime that correspond to particles moving around. If the particle arrangements obey certain principles, they give rise to emergent phenomena that are pretty independent of the particle substrate, and have a totally different feel to them. A great example of this is information processing, in computronium. But we've now taken this idea to another level: *If the information processing itself obeys certain principles, it can give rise to the higher-level emergent phenomenon that we call consciousness.* This places your conscious experience not one but two levels up from the matter. No wonder your mind feels non-physical!

This raises a question: What are these principles that information processing needs to obey to be conscious? I don't pretend to know what conditions are *sufficient* to guarantee consciousness, but here are four *necessary* conditions that I'd bet on and have explored in my research:

Principle	**Definition**
Information principle	A conscious system has substantial information-storage capacity.
Dynamics principle	A conscious system has substantial information-processing capacity.
Independence principle	A conscious system has substantial independence from the rest of the world.
Integration principle	A conscious system cannot consist of nearly independent parts.

As I said, I think that consciousness is the way information feels when being processed in certain ways. This means that to be conscious, a system needs to be able to store and process information, implying the first two principles. Note that the memory doesn't need to last long: I recommend watching this touching video of Clive Wearing, who appears perfectly conscious even though his memories last less than a minute.[21] I think that a conscious system also needs to be fairly independent from the rest of the world, because otherwise it wouldn't subjectively feel that it had any independent existence whatsoever. Finally, I think that the conscious system needs to be integrated into a unified whole, as Giulio Tononi argued, because if it consisted of two independent parts, then they would feel like two separate conscious entities, rather than one. The first three principles imply *autonomy:* that the system is able to retain and process information without much outside interference, hence determining its own future. All four principles together mean that a system is autonomous but its parts aren't.

If these four principles are correct, then we have our work cut out for us: we need to look for mathematically rigorous theories that embody them and test them experimentally. We also need to determine whether additional principles are needed. Regardless of whether IIT is correct or not, researchers should try to develop competing theories and test all available theories with ever better experiments.

Controversies of Consciousness

We've already discussed the perennial controversy about whether consciousness research is unscientific nonsense and a pointless waste of time. In addition, there are recent controversies at the cutting edge of consciousness research—let's explore the ones that I find most enlightening.

Giulio Tononi's IIT has lately drawn not merely praise but also criticism, some of which has been scathing. Scott Aaronson recently had this to say on his blog: "In my opinion, the fact that Integrated Information Theory is wrong—demonstrably wrong, for reasons that go to its core—puts it in something like the top 2% of all mathematical theories of consciousness ever proposed. Almost all competing theories of consciousness, it seems to me, have been so vague, fluffy and malleable that they can only aspire to wrongness."[22] To the credit of both Scott and Giulio, they never came to blows when I watched them debate IIT at a recent New York University workshop, and they politely listened to each other's arguments. Aaronson showed that certain simple networks of logic gates had extremely high integrated information (Φ) and argued that since they clearly weren't conscious, IIT was wrong. Giulio countered that if they were built, they *would* be conscious, and that Scott's assumption to the contrary was anthropocentrically biased, much as if a slaughterhouse owner claimed that animals couldn't be conscious just because they couldn't talk and were very different from humans. My analysis, with which they both agreed, was that they were at odds about whether integration was merely a *necessary* condition for consciousness (which Scott was OK with) or also a *sufficient* condition (which Giulio claimed). The latter is clearly a stronger and more contentious claim, which I hope we can soon test experimentally.[23]

Another controversial IIT claim is that today's computer architectures can't be conscious, because the way their logic gates connect gives very low integration.[24] In other words, if you upload yourself into a future high-powered robot that accurately simulates every single one of your neurons and synapses, then even if this digital clone looks, talks and acts indistinguishably from you, Giulio claims that it will be an unconscious zombie without subjective experience—which would be disappointing if you uploaded yourself in a quest for subjective immortality.* This claim has been challenged by both David Chalmers and AI professor Murray Shanahan by imagining what would happen if you instead gradually replaced the neural circuits in your brain by hypothetical digital hardware perfectly simulating them.[25] Although your *behavior* would be unaffected by the replacement since the simulation is by assumption perfect, your *experience* would change from conscious initially to unconscious at the end, according to Giulio. But how would it feel in between, as ever more got replaced? When the parts of your brain responsible for your con-

* There's potential tension between this claim and the idea that consciousness is substrate-independent, since even though the information processing may be different at the lowest level, it's by definition identical at the higher levels where it determines behavior.

scious experience of the upper half of your visual field were replaced, would you notice that part of your visual scenery was suddenly missing, but that you mysteriously knew what was there nonetheless, as reported by patients with "blindsight"?[26] This would be deeply troubling, because if you can consciously experience any difference, then you can also tell your friends about it when asked—yet by assumption, your behavior can't change. The only logical possibility compatible with the assumptions is that at exactly the same instance that any one thing disappears from your consciousness, your mind is mysteriously altered so as either to make you lie and deny that your experience changed, or to forget that things had been different.

On the other hand, Murray Shanahan admits that the same gradual-replacement critique can be leveled at *any* theory claiming that you can act conscious without being conscious, so you might be tempted to conclude that acting and being conscious are one and the same, and that externally observable behavior is therefore all that matters. But then you'd have fallen into the trap of predicting that you're unconscious while dreaming, even though you know better.

A third IIT controversy is whether a conscious entity can be made of parts that are separately conscious. For example, can society as a whole gain consciousness without the people in it losing theirs? Can a conscious brain have parts that are also conscious on their own? The prediction from IIT is a firm "no," but not everyone is convinced. For example, some patients with lesions severely reducing communication between the two halves of their brain experience "alien hand syndrome," where their right brain makes their left hand do things that the patients claim they aren't causing or understanding—sometimes to the point that they use their other hand to restrain their "alien" hand. How can we be so sure that there aren't two separate consciousnesses in their head, one in the right hemisphere that's unable to speak and another in the left hemisphere that's doing all the talking and claiming to speak for both of them? Imagine using future technology to build a direct communication link between two human brains, and gradually increasing the capacity of this link until communication is as efficient between the brains as it is within them. Would there come a moment when the two individual consciousnesses suddenly disappear and get replaced by a single unified one as IIT predicts, or would the transition be gradual so that the individual consciousnesses coexisted in some form even as a joint experience began to emerge?

Another fascinating controversy is whether experiments underestimate how much we're conscious of. We saw earlier that although we *feel* we're visually conscious of vast amounts of information involving colors, shapes, objects and seemingly everything that's in front of us, experiments have shown that we can only remember and report a dismally small fraction of this.[27] Some researchers have tried to resolve this discrepancy by asking whether we may sometimes have "consciousness without access," that is, subjective experience of things that are too complex to fit into our working memory for later use.[28] For example, when you experience *inattentional blindness* by being too distracted to notice an object in plain sight, this doesn't imply that you had no conscious visual experience of it, merely that it wasn't stored in your working memory.[29] Should it count as forgetfulness rather than blindness? Other researchers reject this idea that people can't be trusted about what they say they experienced, and warn of its implications. Murray Shanahan imagines a clinical trial where patients report complete pain relief thanks to a new wonder drug, which nonetheless gets rejected by a government panel: "The patients only think they are not in pain. Thanks to neuroscience, we know better."[30] On the other hand, there have been cases where patients who accidentally awoke during surgery were given a drug to make them forget the ordeal. Should we trust their subsequent report that they experienced no pain?[31]

How Might AI Consciousness Feel?

If some future AI system is conscious, then what will it subjectively experience? This is the essence of the "even harder problem" of consciousness, and forces us up to the second level of difficulty depicted in figure 8.1. Not only do we currently lack a theory that answers this question, but we're not even sure whether it's logically possible to fully answer it. After all, what could a satisfactory answer sound like? How would you explain to a person born blind what the color red looks like?

Fortunately, our current inability to give a complete answer doesn't prevent us from giving partial answers. Intelligent aliens studying the human sensory system would probably infer that colors are qualia that feel associated with each point on a two-dimensional surface (our visual field), while sounds don't feel as spatially localized, and pains are qualia that feel associated with different parts of our body. From discovering that our retinas have three types of light-sensitive cone cells, they could infer that we experience three primary colors and that all other color qualia result from combining them. By measuring how long it takes neurons to transmit information across the brain, they could conclude that we experience no more than about ten conscious thoughts or perceptions per second, and that when we watch movies on our TV at twenty-four frames per second, we experience this not as a sequence of still images, but as continuous motion. From measuring how fast adrenaline is released into our bloodstream and how long it remains before being broken down, they could predict that we feel bursts of anger starting within seconds and lasting for minutes.

Applying similar physics-based arguments, we can make some educated guesses about certain aspects of how an artificial consciousness may feel. First of all, the space of possible AI experiences is *huge* compared to what we humans can experience. We have one class of qualia for each of our senses, but AIs can have vastly more types of sensors and internal representations of information, so we must avoid the pitfall of assuming that being an AI necessarily feels similar to being a person.

Second, a brain-sized artificial consciousness could have millions of times more experiences than us per second, since electromagnetic signals travel at the speed of light—millions of times faster than neuron signals. However, the larger the AI, the slower its global thoughts must be to allow information time to flow between all its parts, as we saw in chapter 4. We'd therefore expect an Earth-sized "Gaia" AI to have only about ten conscious experiences per second, like a human, and a galaxy-sized AI could have only one global thought every 100,000 years or so—so no more than about a hundred experiences during the entire history of our Universe thus far! This would give large AIs a seemingly irresistible incentive to delegate computations to the smallest subsystems capable of handling them, to speed things up, much like our conscious mind has delegated the blink reflex to a small, fast and unconscious subsystem. Although we saw above that the conscious information processing in our brains appears to be merely the tip of an otherwise unconscious iceberg, we should expect the situation to be even more extreme for large future AIs: if they have a single consciousness, then it's likely to be unaware of almost all the information processing taking place within it. Moreover, although the conscious experiences that it enjoys may be extremely complex, they're also snail-paced compared to the rapid activities of its smaller parts.

This really brings to a head the aforementioned controversy about whether parts of a conscious entity can be conscious too. IIT predicts not, which means that if a future astronomically large AI is conscious, then almost all its information processing is unconscious. This would mean that if a civilization of smaller AIs improves its communication abilities to the point that a single conscious hive mind emerges, their much faster individual consciousnesses are suddenly extinguished. If the IIT prediction is wrong, on the other hand, the hive mind can coexist with the panoply of smaller conscious minds. Indeed, one could even imagine a nested hierarchy of consciousnesses at all levels from microscopic to cosmic.

As we saw above, the unconscious information processing in our human brains appears linked to the effortless, fast and automatic way of thinking that psychologists call "System 1."[32] For example, your System 1 might inform your consciousness that its highly complex analysis of visual input data has determined that your best friend has arrived, without giving you any idea how the computation took place. If this link between systems and consciousness proves to be valid, then it will be tempting to generalize this terminology to AIs, denoting all rapid routine tasks delegated to unconscious subunits as the AI's System 1. The effortful, slow and controlled global thinking of the AI would, if conscious, be the AI's System 2. We humans also have conscious experiences involving what I'll term "System 0": raw passive perception that takes place even when you sit without moving or thinking and merely observe the world around you. Systems 0, 1 and 2 seem progressively more complex, so it's striking that only the middle one appears unconscious. IIT explains this by saying that raw sensory information in System 0 is stored in grid-like brain structures with very high integration, while System 2 has high integration because of feedback loops, where all the information you're aware of right now can affect your future brain states. On the other hand, it was precisely the conscious-grid prediction that triggered Scott Aaronson's aforementioned IIT-critique. In summary, if a theory solving the pretty hard problem of consciousness can one day pass a rigorous battery of experimental tests so that we start taking its predictions seriously, then it will also greatly narrow down the options for the even harder problem of what future conscious AIs may experience.

Some aspects of our subjective experience clearly trace back to our evolutionary origins, for example our emotional desires related to self-preservation (eating, drinking, avoiding getting killed) and reproduction. This means that it should be possible to create AI that never experiences qualia such as hunger, thirst, fear or sexual desire. As we saw in the last chapter, if a highly intelligent AI is programmed to have virtually any sufficiently ambitious goal, it's likely to strive for self-preservation in order to be able to accomplish that goal. If they're part of a society of AIs, however, they might lack our strong human fear of death: as long as they've backed themselves up, all they stand to lose are the memories they've accumulated since their most recent backup, as long as they're confident that their backed-up software will be used. In addition, the ability to readily copy information and software between AIs would probably reduce the strong sense of individuality that's so characteristic of our human consciousness: there would be less of a distinction between you and me if we could easily share and copy all our memories and abilities, so a group of nearby AIs may feel more like a single organism with a hive mind.

Would an artificial consciousness feel that it had free will? Note that, although philosophers have spent millennia quibbling about whether *we* have free will without reaching consensus even on how to define the question,[33] I'm asking a different question, which is arguably easier to tackle. Let me try to persuade you that the answer is simply "Yes, any conscious decision maker will subjectively *feel* that it has free will, regardless of whether it's biological or artificial." Decisions fall on a spectrum between two extremes:

1. You know exactly why you made that particular choice.
2. You have no idea why you made that particular choice—it felt like you chose randomly on a whim.

Free-will discussions usually center around a struggle to reconcile our goal-oriented decision-making behavior with the laws of physics: if you're choosing between the following two explanations for what you did, then which one is correct: "*I asked her on a date because I really liked her*" or "*My particles made me do it by moving according to the laws of physics*"? But we saw in the last chapter that *both* are correct: what feels like goal-oriented behavior can emerge from goal-less deterministic laws of physics. More specifically, when a system (brain or AI) makes a decision of type 1, it computes what to decide using some deterministic algorithm, and the reason it feels like it decided is that it in fact did decide when computing what to do. Moreover, as emphasized by Seth Lloyd,[34] there's a famous computer-science theorem saying that for almost all computations, there's no faster way of determining their outcome than actually running them. This means that it's typically impossible for you to figure out what you'll decide to do in a second in less than a second, which helps reinforce your experience of having free will. In contrast, when a system (brain or AI) makes a decision of type 2, it simply programs its mind to base its decision on the output of some subsystem that acts as a random number generator. In brains and computers, effectively random numbers are easily generated by amplifying noise. Regardless of where on the spectrum from 1 to 2 a decision falls, both biological and artificial consciousnesses therefore feel that they have free will: they feel that it is really they who decide and they can't predict with certainty what the decision will be until they've finished thinking it through.

Some people tell me that they find causality degrading, that it makes their thought processes meaningless and that it renders them "mere" machines. I find such negativity absurd and unwarranted. First of all, there's nothing "mere" about human brains, which, as far as I'm concerned, are the most amazingly sophisticated physical objects in our known Universe. Second, what alternative would they prefer? Don't they want it to be their own thought processes (the computations performed by their brains) that make their decisions?

Their subjective experience of free will is simply how their computations feel from inside: they don't know the outcome of a computation until they've finished it. That's what it means to say that the computation *is* the decision.

Meaning

Let's end by returning to the starting point of this book: How do we want the future of life to be? We saw in the previous chapter how diverse cultures around the globe all seek a future teeming with positive experiences, but that fascinatingly thorny controversies arise when seeking consensus on what should count as positive and how to make trade-offs between what's good for different life forms. But let's not let those controversies distract us from the elephant in the room: there can be no positive experiences if there are no experiences at all, that is, if there's no consciousness. In other words, without consciousness, there can be no happiness, goodness, beauty, meaning or purpose—just an astronomical waste of space. This implies that when people ask about the meaning of life as if it were the job of our cosmos to give meaning to our existence, they're getting it backward: *It's not our Universe giving meaning to conscious beings, but conscious beings giving meaning to our Universe.* So the very first goal on our wish list for the future should be retaining (and hopefully expanding) biological and/or artificial consciousness in our cosmos, rather than driving it extinct.

If we succeed in this endeavor, then how will we humans feel about coexisting with ever smarter machines? Does the seemingly inexorable rise of artificial intelligence bother you and if so, why? In chapter 3, we saw how it should be relatively easy for AI-powered technology to satisfy our basic needs such as security and income as long as the political will to do so exists. However, perhaps you're concerned that being well fed, clad, housed and entertained isn't enough. If we're guaranteed that AI will take care of all our practical needs and desires, might we nonetheless end up feeling that we lack meaning and purpose in our lives, like well-kept zoo animals?

Traditionally, we humans have often founded our self-worth on the idea of *human exceptionalism:* the conviction that we're the smartest entities on the planet and therefore unique and superior. The rise of AI will force us to abandon this and become more humble. But perhaps that's something we should do anyway: after all, clinging to hubristic notions of superiority over others (individuals, ethnic groups, species and so on) has caused awful problems in the past, and may be an idea ready for retirement. Indeed, human exceptionalism hasn't only caused grief in the past, but it also appears unnecessary for human flourishing: if we discover a peaceful extraterrestrial civilization far more advanced than us in science, art and everything else we care about, this presumably wouldn't prevent people from continuing to experience meaning and purpose in their lives. We could retain our families, friends and broader communities, and all activities that give us meaning and purpose, hopefully having lost nothing but arrogance.

As we plan our future, let's consider the meaning not only of our own lives, but also of our Universe itself. Here two of my favorite physicists, Steven Weinberg and Freeman Dyson, represent diametrically opposite views. Weinberg, who won the Nobel Prize for foundational work on the standard model of particle physics, famously said, "The more the universe seems comprehensible, the more it also seems pointless."[35] Dyson, on the other hand, is much more optimistic, as we saw in chapter 6: although he agrees that our Universe *was* pointless, he believes that life is now filling it with ever more meaning, with the best yet to come if life succeeds in spreading throughout the cosmos. He ended his seminal 1979 paper thus: "Is Weinberg's universe or mine closer to the truth? One day, before long, we shall know."[36] If our Universe goes back to being permanently unconscious because we drive Earth life extinct or because we let unconscious zombie AI take over our Universe, then Weinberg will be vindicated in spades.

From this perspective, we see that although we've focused on the future of intelligence in this book, the future of consciousness is even more important, since that's what enables meaning. Philosophers like to go Latin on this distinction, by contrasting *sapience* (the ability to think intelligently) with *sentience* (the ability to subjectively experience qualia). We humans have built our identity on being *Homo sapiens*, the smartest entities around. As we prepare to be humbled by ever smarter machines, I suggest that we rebrand ourselves as *Homo sentiens*!

THE BOTTOM LINE:

- There's no undisputed definition of "consciousness." I use the broad and non-anthropocentric definition *consciousness = subjective experience.*
- Whether AIs are conscious in that sense is what matters for the thorniest ethical and philosophical problems posed by the rise of AI: Can AIs suffer? Should they have rights? Is uploading a subjective suicide? Could a future cosmos teeming with AIs be the ultimate zombie apocalypse?
- The problem of understanding intelligence shouldn't be conflated with three separate problems of consciousness: the "pretty hard problem" of predicting which physical systems are conscious, the "even harder problem" of predicting qualia, and the "*really* hard problem" of why anything at all is conscious.
- The "pretty hard problem" of consciousness is scientific, since a theory that predicts which of your brain processes are conscious is experimentally testable and falsifiable, while it's currently unclear how science could fully resolve the two harder problems.
- Neuroscience experiments suggest that many behaviors and brain regions are unconscious, with much of our conscious experience representing an after-the-fact summary of vastly larger amounts of unconscious information.
- Generalizing consciousness predictions from brains to machines requires a theory. Consciousness appears to require not a particular kind of particle or field, but a particular kind of information processing that's fairly autonomous and integrated, so that the whole system is rather autonomous but its parts aren't.
- Consciousness might feel so non-physical because it's doubly substrate-independent: if consciousness is the way information feels when being processed in certain complex ways, then it's merely the structure of the information processing that matters, not the structure of the matter doing the information processing.
- If artificial consciousness is possible, then the space of possible AI experiences is likely to be huge compared to what we humans can experience, spanning a vast spectrum of qualia and timescales—all sharing a feeling of having free will.
- Since there can be no meaning without consciousness, it's not our Universe giving meaning to conscious beings, but conscious beings giving meaning to our Universe.
- This suggests that as we humans prepare to be humbled by ever smarter machines, we take comfort mainly in being *Homo sentiens*, not *Homo sapiens*.

Epilogue

The Tale of the FLI Team

The saddest aspect of life right now is that science gathers knowledge faster than society gathers wisdom.

Isaac Asimov

Here we are, my dear reader, at the end of the book, after exploring the origin and fate of intelligence, goals and meaning. So how can we translate these ideas into action? What concretely should we *do* to make our future as good as possible? This is precisely the question I'm asking myself right now as I sit here in my window seat en route from San Francisco back to Boston on January 9, 2017, from the AI conference we just organized in Asilomar, so let me end this book by sharing my thoughts with you.

Meia is catching up on sleep next to me after the many short nights of preparing and organizing. Wow—what a wild week it's been! We managed to bring almost all the people I've mentioned in this book together for a few days to this Puerto Rico sequel, including entrepreneurs such as Elon Musk and Larry Page and AI research leaders from academia and companies such as DeepMind, Google, Facebook, Apple, IBM, Microsoft and Baidu, as well as economists, legal scholars, philosophers and other amazing thinkers (see figure 9.1). The results superseded even my high expectations, and I'm feeling more optimistic about the future of life than I have in a long time. In this epilogue, I'm going to tell you why.

FLI Is Born

Ever since I learned about the nuclear arms race at age fourteen, I've been concerned that the power of our technology was growing faster than the wisdom with which we manage it. I therefore decided to sneak a chapter about this challenge into my first book, *Our Mathematical Universe*, even though the rest of it was primarily about physics. I made a New Year's resolution for 2014 that I was no longer allowed to complain about anything without putting some serious thought into what I could personally do about it, and I kept my pledge during my book tour that January: Meia and I did lots of brainstorming about starting some sort of nonprofit organization focused on improving the future of life through technological stewardship.

She insisted that we give it a positive name as different as possible from "Doom & Gloom Institute" and "Let's-Worry-about-the-Future Institute." Since Future of Humanity Institute was already taken, we converged on the Future of Life Institute (FLI), which had the added advantage of being more inclusive. On January 22, the book tour took us to Santa Cruz, and as the California Sun set over the Pacific, we enjoyed dinner with our old friend Anthony Aguirre and persuaded him to join forces with us. He's not only one of the wisest and most idealistic people I know, but also someone who's managed to put up with running another nonprofit organization, the *Foundational Questions Institute* (see http://fqxi.org), with me for over a decade.

The following week, the tour took me to London. Since the future of AI was very much on my mind, I reached out to Demis Hassabis, who graciously invited me to visit DeepMind's headquarters. I was awestruck by how much they'd grown since he visited me at MIT two years earlier. Google had just bought them for about $650 million, and seeing their vast office landscape filled with brilliant minds pursuing Demis' audacious goal to "solve intelligence" gave me a visceral feeling that success was a real possibility.

The next evening, I spoke with my friend Jaan Tallinn using Skype, the software he'd helped create. I explained our FLI vision, and an hour later, he'd decided to take a chance on us, funding us at up to

Figure 9.1: Our January 2017 Asilomar conference, the Puerto Rico sequel, brought together a remarkable group of researchers in AI and related fields. Back row, from left to right: Patrick Lin, Daniel Weld, Ariel Conn, Nancy Chang, Tom Mitchell, Ray Kurzweil, Daniel Dewey, Margaret Boden, Peter Norvig, Nick Hay, Moshe Vardi, Scott Siskind, Nick Bostrom, Francesca Rossi, Shane Legg, Manuela Veloso, David Marble, Katja Grace, Irakli Beridze, Marty Tenenbaum, Gill Pratt, Martin Rees, Joshua Greene, Matt Scherer, Angela Kane, Amara Angelica, Jeff Mohr, Mustafa Suleyman, Steve Omohundro, Kate Crawford, Vitalik Buterin, Yutaka Matsuo, Stefano Ermon, Michael Wellman, Bas Steunebrink, Wendell Wallach, Allan Dafoe, Toby Ord, Thomas Dietterich, Daniel Kahneman, Dario Amodei, Eric Drexler, Tomaso Poggio, Eric Schmidt, Pedro Ortega, David Leake, Seán Ó hÉigeartaigh, Owain Evans, Jaan Tallinn, Anca Dragan, Sean Legassick, Toby Walsh, Peter Asaro, Kay Firth-Butterfield, Philip Sabes, Paul Merolla, Bart Selman, Tucker Davey, ?, Jacob Steinhardt, Moshe Looks, Josh Tenenbaum, Tom Gruber, Andrew Ng, Kareem Ayoub, Craig Calhoun, Percy Liang, Helen Toner, David Chalmers, Richard Sutton, Claudia Passos-Ferriera, János Krámar, William MacAskill, Eliezer Yudkowsky, Brian Ziebart, Huw Price, Carl Shulman, Neil Lawrence, Richard Mallah, Jurgen Schmidhuber, Dileep George, Jonathan Rothberg, Noah Rothberg. Front row: Anthony Aguirre, Sonia Sachs, Lucas Perry, Jeffrey Sachs, Vincent Conitzer, Steve Goose, Victoria Krakovna, Owen Cotton-Barratt, Daniela Rus, Dylan Hadfield-Menell, Verity Harding, Shivon Zilis, Laurent Orseau, Ramana Kumar, Nate Soares, Andrew McAfee, Jack Clark, Anna Salamon, Long Ouyang, Andrew Critch, Paul Christiano, Yoshua Bengio, David Sanford, Catherine Olsson, Jessica Taylor, Martina Kunz, Kristinn Thorisson, Stuart Armstrong, Yann LeCun, Alexander Tamas, Roman Yampolskiy, Marin Soljačić, Lawrence Krauss, Stuart Russell, Eric Brynjolfsson, Ryan Calo, ShaoLan Hsueh, Meia Chita-Tegmark, Kent Walker, Heather Roff, Meredith Whittaker, Max Tegmark, Adrian Weller, Jose Hernandez-Orallo, Andrew Maynard, John Hering, Abram Demski, Nicolas Berggruen, Gregory Bonnet, Sam Harris, Tim Hwang, Andrew Snyder-Beattie, Marta Halina, Sebastian Farquhar, Stephen Cave, Jan Leike, Tasha McCauley, Joseph Gordon-Levitt. Arrived later: Guru Banavar, Demis Hassabis, Rao Kambhampati, Elon Musk, Larry Page, Anthony Romero.

$100,000 a year! Few things touch me more than when someone places more trust in me than I've earned, so it meant the world to me when a year later, after the Puerto Rico conference I mentioned in chapter 1, he joked that this was the best investment he'd ever made.

The next day, my publisher had left a gap in my schedule, which I filled with a visit to the London Science Museum. After having obsessed about the past and future of intelligence for so long, I suddenly felt that I was walking through a physical manifestation of my thoughts. They'd assembled a fantastic collection of stuff representing our growth of knowledge, from Stephenson's Rocket locomotive to the Model T Ford, a life-size Apollo 11 lunar lander replica and computers dating all the way from Babbage's "Difference Engine" mechanical calculator to present-day hardware. They also had an exhibit about the history of our understanding of the mind, from Galvano's frog-leg experiments to neurons, EEG and fMRI.

I very rarely cry, but that's what I did on the way out—and in a tunnel full of pedestrians, no less, en route to the South Kensington tube station. Here were all these people going about their lives blissfully unaware of what I was thinking. First we humans discovered how to replicate some natural processes with machines, making our own wind and lightning, and our own mechanical horsepower. Gradually, we started realizing that our bodies were also machines. Then the discovery of nerve cells started blurring the borderline between body and mind. Then we started building machines that could outperform not only our muscles, but our minds as well. So in parallel with discovering what we are, are we inevitably making ourselves obsolete? That would be poetically tragic.

This thought scared me, but it also strengthened my resolve to keep my New Year's resolution. I felt that we needed one more person to complete our team of FLI founders, who'd spearhead a team of idealistic young volunteers. The logical choice was Viktoriya Krakovna, a brilliant Harvard grad student who'd not only won a silver medal in the International Mathematics Olympiad, but also founded the Citadel, a house for about a dozen young idealists who wanted reason to play a greater role in their lives and the world. Meia and I invited her over to our place five days later to tell her about our vision, and before we'd finished the sushi, FLI had been born.

The Puerto Rico Adventure

This marked the beginning of an amazing adventure, which still continues. As I mentioned in chapter 1, we held regular brainstorming meetings at our house with dozens of idealistic students, professors

Figure 9.2: Jaan Tallinn, Anthony Aguirre, yours truly, Meia Chita-Tegmark and Viktoriya Krakovna celebrate our incorporation of FLI with sushi on May 23, 2014.

and other local thinkers, where the top-rated ideas transformed into projects—the first being that AI op-ed from chapter 1 with Stephen Hawking, Stuart Russell and Frank Wilczek that helped ignite the public debate. In parallel with the baby steps of setting up a new organization (such as incorporating, recruiting an advisory board and launching a website), we held a fun launch event in front of a packed MIT auditorium, at which Alan Alda explored the future of technology with leading experts.

We focused the rest of the year on pulling together the Puerto Rico conference which, as I mentioned in chapter 1, aimed to engage the world's leading AI researchers in the discussion of how to keep AI beneficial. Our goal was to shift the AI-safety conversation from worrying to working: from bickering about how worried to be, to agreeing on concrete research projects that could be started right away to maximize the chance of a good outcome. To prepare, we collected promising AI-safety research ideas from around the world and sought community feedback on our growing project list. With the help of Stuart Russell and a group of hardworking young volunteers, especially Daniel Dewey, János Krámar and Richard Mallah, we distilled these research priorities into a document to be discussed at the conference.[1] Building consensus that there was lots of valuable AI-safety research to be done would, we hoped, encourage people to start doing such research. The ultimate moonshot triumph would be if it could even persuade someone to fund it since, so far, there had been essentially no support for such work from government funding agencies.

Enter Elon Musk. On August 2, he appeared on our radar by famously tweeting "Worth reading Superintelligence by Bostrom. We need to be super careful with AI. Potentially more dangerous than nukes." I reached out to him about our efforts, and got to speak with him by phone a few weeks later. Although I felt quite nervous and starstruck, the outcome was outstanding: he agreed to join our FLI scientific advisory board, to attend our conference and potentially to fund a first-ever AI-safety research program to be announced in Puerto Rico. This electrified all of us at FLI, and made us redouble our efforts to create an awesome conference, identify promising research topics and build community support for them.

I finally got to meet Elon in person for further planning when he came to MIT two months later for a space symposium. It felt very strange to be alone with him in a small green room just moments after he'd enraptured over a thousand MIT students like a rock star, but after a few minutes, all I could think of was our joint project. I instantly liked him. He radiated sincerity, and I was inspired by how much he cared about the long-term future of humanity—and how he audaciously turned his aspiration into actions. He wanted humanity to explore and settle our Universe, so he started a space company. He wanted sustainable energy, so he started a solar company and an electric-car company. Tall, handsome, eloquent and incredibly knowledgeable, it was easy to understand why people listened to him.

Unfortunately, this MIT event also taught me how fear-driven and divisive media can be. Elon's stage performance consisted of an hour of fascinating discussion about space exploration, which I think would have made great TV. At the very end, a student asked him an off-topic question about AI. His answer included the phrase "with artificial intelligence, we are summoning the demon," which became the *only* thing that most media reported—and generally out of context. It struck me that many journalists were inadvertently doing the *exact opposite* of what we were trying to accomplish in Puerto Rico. Whereas we wanted to build community consensus by highlighting the common ground, the media had an incentive to highlight the divisions. The more controversy they could report, the greater their Nielsen ratings and ad revenue. Moreover, whereas we wanted to help people from across the spectrum of opinions to come together, get along and understand each other better, media coverage inadvertently made people across the opinion spectrum upset at one another, fueling misunderstandings by publishing only their most provocative-sounding quotes without context. For this reason, we decided to ban journalists from the Puerto Rico meeting and impose the "Chatham House Rule," which prohibits participants from subsequently revealing who said what.*

* This experience also made me rethink how I personally should interpret news. Although I'd obviously been aware that most outlets have their own political agenda, I now realized that they also have a bias away from the center on all issues, even nonpolitical ones.

Although our Puerto Rico conference ended up being a success, it didn't come easy. The countdown mostly required diligent prep work, for example me phoning or skyping large numbers of AI researchers to assemble a critical mass of participants to attract the other attendees, and there were also dramatic moments—such as when I got up by 7 a.m. on December 27 to reach Elon on a lousy phone connection to Uruguay, and was told "I don't think this is gonna work." He was concerned that an AI-safety research program might provide a false sense of security, enabling reckless researchers to forge ahead while paying lip service to safety. But then, despite the sound incessantly cutting out, we extensively talked through the huge benefits of mainstreaming the topic and getting more AI researchers working on AI safety. After the call dropped, he sent me one of my favorite emails ever: "Lost the call at the end there. Anyway, docs look fine. I'm happy to support the research with $5M over three years. Maybe we should make it $10M?"

Four days later, 2015 got off to a good start for Meia and me as we briefly relaxed before the meeting, dancing in the new year on a Puerto Rico beach illuminated by fireworks. The conference got off to a great start too: there was remarkable consensus that more AI-safety research was needed, and based on further input from the conference participants, that research priorities document we'd worked so hard on was improved and finalized. We passed around that safety-research-endorsing open letter from chapter 1, and were delighted that almost everyone signed it.

Meia and I had a magical meeting with Elon in our hotel room where he blessed the detailed plans for our grants program. She was touched by how down-to-earth and candid he was about his personal life, and how much interest he took in us. He asked us how we met, and liked Meia's elaborate story. The next day, we filmed an interview with him about AI safety and why he wanted to support it and everything seemed on track.[2]

The conference climax, Elon's donation announcement, was scheduled for 7 p.m. on Sunday, January 4, 2015, and I'd been so tense about it that I'd tossed and turned in my sleep the night before. And then, just fifteen minutes before we were supposed to head to the session where it would happen, we hit a snag! Elon's assistant called and said that it looked like Elon might not be able to go through with the announcement, and Meia said she'd never seen me look more stressed or disappointed. Elon finally came by, and I could hear the seconds counting down to the session start as we sat there and talked. He explained that they were just two days away from a crucial SpaceX rocket launch where they hoped to pull off the first-ever successful landing of the first stage on a drone ship, and that since this was a huge milestone, the SpaceX team didn't want to distract from it with concurrent media splashes involving him. Anthony Aguirre, cool and levelheaded as always, pointed out that this meant that *nobody* wanted media attention for this, neither Elon nor the AI community. We arrived a few minutes late to the session I was moderating, but we had a plan: no dollar amount would get mentioned, to ensure that the announcement wasn't newsworthy, and I'd lord Chatham House over everyone to keep Elon's announcement secret from the world for nine days if his rocket reached the space station, regardless of whether the landing succeeded; he said he'd need even more time if the rocket exploded on launch.

The countdown to the announcement finally reached zero. The superintelligence panelists that I'd moderated still sat there next to me onstage in their chairs: Eliezer Yudkowsky, Elon Musk, Nick Bostrom, Richard Mallah, Murray Shanahan, Bart Selman, Shane Legg and Vernor Vinge. People gradually stopped applauding, but the panelists remained seated, because I'd told them to stay without explaining why. Meia later told me that her pulse reached the stratosphere around now, and that she clutched Viktoriya Krakovna's calming hand under the table. I smiled, knowing that this was the moment we'd worked, hoped and waited for.

I was very happy that there was such consensus at the meeting that more research was needed for keeping AI beneficial, I said, and that there were so many concrete research directions we could work on right away. But there had been talk of serious risks in this session, I added, so it would be nice to raise our spirits and get into an upbeat mood before heading out to the bar and the conference banquet that had been set up outside. "And I'm therefore giving the microphone to . . . Elon Musk!" I felt that history was in the making as Elon took the mic and announced that he would donate a large amount of money to AI-safety research. Unsurprisingly, he brought down the house. As planned, he didn't mention how much, but I knew that it was a cool $10 million, as we'd agreed.

Meia and I went to visit our parents in Sweden and Romania after the conference, and with bated breath, we watched the live-streamed rocket launch with my dad in Stockholm. The landing attempt unfortunately ended with what Elon euphemistically calls an RUD, "rapid unscheduled disassembly," and pulling off a successful ocean landing took his team another fifteen months.[3] However, all the satellites were successfully launched into orbit, as was our grants program via a tweet by Elon to his millions of followers.[4]

Mainstreaming AI Safety

A key goal of the Puerto Rico conference had been to mainstream AI-safety research, and it was exhilarating to see this unfold in multiple steps. First there was the meeting itself, where many researchers started feeling comfortable engaging with the topic once they realized that they were part of a growing community of peers. I was deeply touched by encouragement from many participants. For example, Cornell University AI professor Bart Selman emailed me saying, "I've honestly never seen a better organized or more exciting and intellectually stimulating scientific meeting."

The next mainstreaming step began on January 11 when Elon tweeted "World's top artificial intelligence developers sign open letter calling for AI-safety research,"[5] linking to a sign-up page that soon racked up over eight thousand signatures, including many of the world's most prominent AI builders. It suddenly became harder to claim that people concerned about AI safety didn't know what they were talking about, because this now implied that a who's who of leading AI researchers didn't know what they were talking about. The open letter was reported by media around the world in a way that made us grateful that we'd barred journalists from our conference. Although the most alarmist word in the letter was "pitfalls," it nonetheless triggered headlines such as "Elon Musk and Stephen

Hawking Sign Open Letter in Hopes of Preventing Robot Uprising," illustrated by murderous terminators. Of the hundreds of articles we spotted, our favorite was one mocking the others, writing that "a headline that conjures visions of skeletal androids stomping human skulls underfoot turns complex, transformative technology into a carnival sideshow."[6] Fortunately, there were also many sober news articles, and they gave us another challenge: keeping up with the torrent of new signatures, which needed to be manually verified to protect our credibility and weed out pranks such as "HAL 9000," "Terminator," "Sarah Jeanette Connor" and "Skynet." For this and our future open letters, Viktoriya Krakovna and János Krámar helped organize a volunteer brigade of checkers that included Jesse Galef, Eric Gastfriend and Revathi Vinoth Kumar working in shifts, so that when Revathi went to sleep in India, she passed the baton to Eric in Boston, and so on.

The third mainstreaming step began four days later, when Elon tweeted a link to our announcement that he was donating $10 million to AI-safety research.[7] A week later, we launched an online portal where researchers from around the world could apply and compete for this funding. We were able to whip the application system together so quickly only because Anthony and I had spent the previous decade running similar competitions for physics grants. The Open Philanthropy Project, a California-based charity focused on high-impact giving, generously agreed to top up Elon's gift to allow us to give more grants. We weren't sure how many applicants we'd get, since the topic was novel and the deadline was short. The response blew us away, with about three hundred teams from around the world asking for about $100 million. A panel of AI professors and other researchers carefully reviewed the proposals and selected thirty-seven winning teams, who were funded for up to three years. When we announced the list of winners, it marked the first time that the media response to our activities was fairly nuanced and free of killer-robot pictures. It was finally sinking in that AI safety wasn't empty talk: there was actual useful work to be done, and lots of great research teams were rolling up their sleeves to join the effort.

The fourth mainstreaming step happened organically over the

next two years, with scores of technical publications and dozens of workshops on AI safety around the world, typically as parts of mainstream AI conferences. Persistent people had tried for many years to engage the AI community in safety research, with limited success, but now things really took off. Many of these publications were funded by our grants program and we at FLI did our best to help organize and fund as many of these workshops as we could, but a growing fraction of them were enabled by AI researchers investing their own time and resources. As a result, ever more researchers learned about safety research from their own colleagues, discovering that aside from being useful, it could also be fun, involving interesting mathematical and computational problems to puzzle over.

Complicated equations aren't everyone's idea of fun, of course. Two years after our Puerto Rico conference, we preceded our Asilomar conference with a technical workshop where our FLI grant winners could showcase their research, and watched slide after slide with mathematical symbols on the big screen. Moshe Vardi, an AI professor at Rice University, joked that he knew we'd succeeded in establishing an AI-safety research field once the meetings got boring.

This dramatic growth of AI-safety work wasn't limited to academia. Amazon, DeepMind, Facebook, Google, IBM and Microsoft launched an industry partnership for beneficial AI.[8] Major new AI-safety donations enabled expanded research at our largest nonprofit sister organizations: the Machine Intelligence Research Institute in Berkeley, the Future of Humanity Institute in Oxford and the Centre for the Study of Existential Risk in Cambridge (UK). Further donations of $10 million or more kick-started additional beneficial-AI efforts: the Leverhulme Centre for the Future of Intelligence in Cambridge, the K&L Gates Endowment for Ethics and Computational Technologies in Pittsburgh and the Ethics and Governance of Artificial Intelligence Fund in Miami. Last but not least, with a billion-dollar commitment, Elon Musk partnered with other entrepreneurs to launch OpenAI, a nonprofit company in San Francisco pursuing beneficial AI. AI-safety research was here to stay.

In lockstep with this surge of research came a surge of opinions being expressed, both individually and collectively. The industry

Partnership on AI published its founding tenets, and long reports with lists of recommendations were published by the U.S. government, Stanford University and the IEEE (the world's largest organization of technical professionals), together with dozens of other reports and position papers from elsewhere.[9]

We were eager to facilitate meaningful discussion among the Asilomar attendees and learn what, if anything, this diverse community agreed on. Lucas Perry therefore took on the heroic task of reading all of those documents we'd found and extracting all their opinions. In a marathon effort initiated by Anthony Aguirre and concluded by a series of long telecons, our FLI team then attempted to group similar opinions together and strip away redundant bureaucratic verbiage to end up with a single list of succinct principles, also including unpublished but influential opinions that had been expressed more informally in talks and elsewhere. But this list still included plenty of ambiguity, contradiction and room for interpretation, so the month before the conference, we shared it with the participants and collected their opinions and suggestions for improved or novel principles. This community input produced a significantly revised principle list for use at the conference.

In Asilomar, the list was further improved in two steps. First, small groups discussed the principles they were most interested in (figure

Figure 9.3: Groups of great minds ponder AI principles in Asilomar.

9.4), producing detailed refinements, feedback, new principles and competing versions of old ones. Finally, we surveyed all attendees to determine the level of support for each version of each principle.

This collective process was both exhaustive and exhausting, with Anthony, Meia and I curtailing sleep and lunch time at the conference in our scramble to compile everything needed in time for the next steps. But it was also exciting. After such detailed, thorny and sometimes contentious discussions and such a wide range of feedback, we were astonished by the high level of consensus that emerged around many of the principles during that final survey, with some getting over 97% support. This consensus allowed us to set a high bar for inclusion in the final list: we kept only principles that at least 90% of the attendees agreed on. Although this meant that some popular principles were dropped at the last minute, including some of my personal favorites,[10] it enabled most of the participants to feel comfortable endorsing all of them on the sign-up sheet that we passed around the auditorium. Here's the result.

The Asilomar AI Principles

Artificial intelligence has already provided beneficial tools that are used every day by people around the world. Its continued development, guided by the following principles, will offer amazing opportunities to help and empower people in the decades and centuries ahead.

Research Issues

§1 Research Goal: *The goal of AI research should be to create not undirected intelligence, but beneficial intelligence.*

§2 Research Funding: *Investments in AI should be accompanied by funding for research on ensuring its beneficial use, including thorny questions in computer science, economics, law, ethics, and social studies, such as:*

- (a) *How can we make future AI systems highly robust, so that they do what we want without malfunctioning or getting hacked?*
- (b) *How can we grow our prosperity through automation while maintaining people's resources and purpose?*

(c) *How can we update our legal systems to be more fair and efficient, to keep pace with AI, and to manage the risks associated with AI?*

(d) *What set of values should AI be aligned with, and what legal and ethical status should it have?*

§3 Science-Policy Link: *There should be constructive and healthy exchange between AI researchers and policy-makers.*

§4 Research Culture: *A culture of cooperation, trust, and transparency should be fostered among researchers and developers of AI.*

§5 Race Avoidance: *Teams developing AI systems should actively cooperate to avoid corner-cutting on safety standards.*

Ethics and Values

§6 Safety: *AI systems should be safe and secure throughout their operational lifetime, and verifiably so where applicable and feasible.*

§7 Failure Transparency: *If an AI system causes harm, it should be possible to ascertain why.*

§8 Judicial Transparency: *Any involvement by an autonomous system in judicial decision-making should provide a satisfactory explanation auditable by a competent human authority.*

§9 Responsibility: *Designers and builders of advanced AI systems are stakeholders in the moral implications of their use, misuse, and actions, with a responsibility and opportunity to shape those implications.*

§10 Value Alignment: *Highly autonomous AI systems should be designed so that their goals and behaviors can be assured to align with human values throughout their operation.*

§11 Human Values: *AI systems should be designed and operated so as to be compatible with ideals of human dignity, rights, freedoms, and cultural diversity.*

§12 Personal Privacy: *People should have the right to access, manage, and control the data they generate, given AI systems' power to analyze and utilize that data.*

§13 Liberty and Privacy: *The application of AI to personal data must not unreasonably curtail people's real or perceived liberty.*

§14 Shared Benefit: *AI technologies should benefit and empower as many people as possible.*

§15 Shared Prosperity: *The economic prosperity created by AI should be shared broadly, to benefit all of humanity.*

§16 Human Control: *Humans should choose how and whether to delegate decisions to AI systems, to accomplish human-chosen objectives.*

§17 Non-subversion: *The power conferred by control of highly advanced AI systems should respect and improve, rather than subvert, the social and civic processes on which the health of society depends.*

§18 AI Arms Race: *An arms race in lethal autonomous weapons should be avoided.*

LONGER-TERM ISSUES

§19 Capability Caution: *There being no consensus, we should avoid strong assumptions regarding upper limits on future AI capabilities.*

§20 Importance: *Advanced AI could represent a profound change in the history of life on Earth, and should be planned for and managed with commensurate care and resources.*

§21 Risks: *Risks posed by AI systems, especially catastrophic or existential risks, must be subject to planning and mitigation efforts commensurate with their expected impact.*

§22 Recursive Self-Improvement: *AI systems designed to recursively self-improve or self-replicate in a manner that could lead to rapidly increasing quality or quantity must be subject to strict safety and control measures.*

§23 Common Good: *Superintelligence should only be developed in the service of widely shared ethical ideals, and for the benefit of all humanity rather than one state or organization.*

The signature list grew dramatically after we posted the principles online, and by now it includes an amazing list of more than a thousand AI researchers and many other top thinkers. If you too want to join as a signatory, you can do so here: http://futureoflife.org/ai-principles.

We were struck not only by the level of consensus about the principles, but also by how strong they were. Sure, some of them sound about as controversial as "Peace, love and motherhood are valuable" at first glance. But many of them have real teeth, as is most easily seen by formulating negations of them. For example, "Superintelligence is impossible!" violates §19, and "It's a total waste to do research on reducing existential risk from AI!" violates §21.

Indeed, as you can see for yourself if you watch our long-term panel discussion on YouTube,[11] Elon Musk, Stuart Russell, Ray Kurzweil, Demis Hassabis, Sam Harris, Nick Bostrom, David Chalmers, Bart Selman and Jaan Tallinn all agreed that superintelligence would probably be developed and that safety research was important.

Mindful Optimism

As I confessed in the opening of this epilogue, I'm feeling more optimistic about the future of life than I have in a long time. I shared my personal story to explain why.

My experiences over the past few years have increased my optimism for two separate reasons. First, I've witnessed the AI community come together in a remarkable way to constructively take on the challenges ahead, often in collaboration with thinkers from other fields. Elon told me after the Asilomar meeting that he found it amazing how AI safety has gone from a fringe issue to mainstream in only a few years, and I'm just as amazed myself. And now it's not merely the near-term issues from chapter 3 that are becoming respectable discussion topics, but even superintelligence and existential risk, as in the Asilomar AI Principles. There's no way that those principles could have been adopted in Puerto Rico two years earlier, where the most scary-sounding word that made it into the open letter was "pitfalls."

I like people-watching, and at one point during the final morning of the Asilomar conference, I stood at the side of the auditorium and watched the participants listen to a discussion about AI and law. To my surprise, a warm and fuzzy feeling swept through me and I suddenly felt very moved. This felt so different from Puerto Rico! Back then, I remember viewing most of the AI community with a combination of respect and fear—not exactly as an opposing team, but as a group that my AI-concerned colleagues and I felt we needed to persuade. But now it felt so obvious that we were all on the *same* team. As you've probably gleaned from reading this book, I still don't have the

Figure 9.4: A growing community searches for answers together in Asilomar.

answers for how to create a great future with AI, so it feels great to be part of a growing community searching for the answers together.

The second reason I've grown more optimistic is that the FLI experience has been empowering. What had triggered my London tears was a feeling of inevitability: that a disturbing future may be coming and there was nothing we could do about it. But the next three years dissolved my fatalistic gloom. If even a ragtag bunch of unpaid volunteers could make a positive difference for what's arguably the most important conversation of our time, then imagine what we can all do if we work together!

Erik Brynjolfsson spoke of two kinds of optimism in his Asilomar talk. First there's the unconditional kind, such as the positive expectation that the Sun will rise tomorrow morning. Then there's what he called "mindful optimism," which is the expectation that good things will happen if you plan carefully and work hard for them. That's the kind of optimism I now feel about the future of life.

So what can *you* do to make a positive difference for the future of life as we enter the age of AI? For reasons I'll soon explain, I think that a great first step is working on becoming a mindful optimist, if you aren't one already. To be a successful mindful optimist, it's crucial to develop positive visions for the future. When MIT students come to my office for career advice, I usually start by asking them where they see themselves in a decade. If a student were to reply "Perhaps I'll be in a cancer ward, or in a cemetery after getting hit by a bus," I'd give her a hard time. Envisioning only negative futures is a terrible approach to career planning! Devoting 100% of one's efforts to avoiding diseases and accidents is a great recipe for hypochondria and paranoia, not happiness. Instead, I'd like to hear her describe her goals with enthusiasm, after which we can discuss strategies for getting there while avoiding pitfalls.

Erik pointed out that according to game theory, positive visions form the foundation of a large fraction of all collaboration in the world, from marriages and corporate mergers to the decision of independent states to form the USA. After all, why sacrifice something you have if you can't imagine the even greater gain that this will provide? This means that we should be imagining positive futures not only for ourselves, but also for society and for humanity itself. In other words, we need more existential hope! Yet, as Meia likes to remind me, from Frankenstein to the Terminator, futuristic visions in literature and film are predominantly dystopian. In other words, we as a society are planning our future about as poorly as that hypothetical MIT student. This is why we need more mindful optimists. And this is why I've encouraged you throughout this book to think about what sort of future you *want* rather than merely what sort of future you *fear*, so that we can find shared goals to plan and work for.

We've seen throughout this book how AI is likely to give us both grand opportunities and tough challenges. A strategy that's likely to help with essentially all AI challenges is for us to get our act together and improve our human society *before* AI fully takes off. We're better off educating our young to make technology robust and beneficial before ceding great power to it. We're better off modernizing our laws before technology makes them obsolete. We're better off resolving international conflicts before they escalate into an arms race in autonomous weapons. We're better off creating an economy that ensures prosperity for all before AI potentially amplifies inequalities. We're better off in a society where AI-safety research results get implemented rather than ignored. And looking further ahead, to challenges related to superhuman AGI, we're better off agreeing on at least some basic ethical standards before we start teaching these standards to powerful machines. In a polarized and chaotic world, people with the power to use AI for malicious purposes will have more motivation and ability to do so, and teams racing to build AGI will feel more pressure to cut corners on safety than to cooperate. In summary, if we can create a more harmonious human society characterized by cooperation toward shared goals, this will improve the prospects of the AI revolution ending well.

In other words, one of the best ways for you to improve the future of life is to improve tomorrow. You have power to do so in many ways. Of course you can vote at the ballot box and tell your politicians what you think about education, privacy, lethal autonomous weapons, technological unemployment and other issues. But you also vote every day through what you choose to buy, what news you choose to consume, what you choose to share and what sort of role model you choose to be. Do you want to be someone who interrupts all their conversations by checking their smartphone, or someone who feels empowered by using technology in a planned and deliberate way? Do you want to own your technology or do you want your technology to own you? What do you want it to mean to be human in the age of AI? Please discuss all this with those around you—it's not only an important conversation, but a fascinating one.

We're the guardians of the future of life now as we shape the age of AI. Although I cried in London, I now feel that there's nothing inevitable about this future, and I know that it's much easier to make a difference than I thought. Our future isn't written in stone and just waiting to happen to us—it's ours to create. Let's create an inspiring one together!

Notes

Chapter 1

1. "The AI Revolution: Our Immortality or Extinction?" *Wait But Why* (January 27, 2015), at http://waitbutwhy.com/2015/01/artificial-intelligence-revolution-2.html.
2. This open letter, "Research Priorities for Robust and Beneficial Artificial Intelligence," can be found at http://futureoflife.org/ai-open-letter/.
3. Example of classic robot alarmism in the media: Ellie Zolfagharifard, "Artificial Intelligence 'Could Be the Worst Thing to Happen to Humanity,'" *Daily Mail*, May 2, 2014; http://tinyurl.com/hawkingbots.

Chapter 2

1. Notes on the origin of the term AGI: http://wp.goertzel.org/who-coined-the-term-agi.
2. Hans Moravec, "When Will Computer Hardware Match the Human Brain?" *Journal of Evolution and Technology* (1998), vol. 1.
3. In the figure showing computing power versus year, the pre-2011 data is from Ray Kurzweil's book *How to Create a Mind*, and subsequent data is computed from the references in https://en.wikipedia.org/wiki/FLOPS.
4. Quantum computing pioneer David Deutsch describes how he views quantum computation as evidence of parallel universes in his *The Fabric of Reality: The Science of Parallel Universes—and Its Implications* (London: Allen Lane, 1997). If you want my own take on quantum parallel universes as the third of four multiverse levels, you'll find it in my previous book: Max Tegmark, *Our Mathematical Universe: My Quest for the Ultimate Nature of Reality* (New York: Knopf, 2014).

Chapter 3

1. Watch "Google DeepMind's Deep Q-learning Playing Atari Breakout" on YouTube at https://tinyurl.com/atariai.

2. See Volodymyr Mnih et al., "Human-Level Control Through Deep Reinforcement Learning," *Nature* 518 (February 26, 2015): 529–533, available online at http://tinyurl.com/ataripaper.
3. Here's a video of the Big Dog robot in action: https://www.youtube.com/watch?v=W1czBcnX1Ww.
4. For reactions to the sensationally creative line 5 move by AlphaGO, see "Move 37!! Lee Sedol vs AlphaGo Match 2," at https://www.youtube.com/watch?v=JNrXgpSEEIE.
5. Demis Hassabis describing reactions to AlphaGo from human Go players: https://www.youtube.com/watch?v=otJKzpNWZT4.
6. For recent improvements in machine translation, see Gideon Lewis-Kraus, "The Great A.I. Awakening," *New York Times Magazine* (December 14, 2016), available online at http://www.nytimes.com/2016/12/14/magazine/the-great-ai-awakening.html. GoogleTranslate is available here at https://translate.google.com.
7. Winograd Schema Challenge competition: http://tinyurl.com/winogradchallenge.
8. Ariane 5 explosion video: https://www.youtube.com/watch?v=qnHn8W1Em6E.
9. Ariane 5 Flight 501 Failure report by the inquiry board: http://tinyurl.com/arianeflop.
10. NASA's Mars Climate Orbiter Mishap Investigation Board Phase I report: http://tinyurl.com/marsflop.
11. The most detailed and consistent account of what caused the Mariner 1 Venus mission failure was incorrect hand-transcription of a single mathematical symbol (a missing overbar): http://tinyurl.com/marinerflop.
12. A detailed description of the failure of the Soviet Phobos 1 Mars mission can be found in Wesley T. Huntress Jr. and Mikhail Ya. Marov, *Soviet Robots in the Solar System* (New York: Praxis Publishing, 2011), p. 308.
13. How unverified software cost Knight Capital $440 million in 45 minutes: http://tinyurl.com/knightflop1 and http://tinyurl.com/knightflop2.
14. U.S. government report on the Wall Street "flash crash": "Findings Regarding the Market Events of May 6, 2010" (September 30, 2010), at http://tinyurl.com/flashcrashreport.
15. 3-D printing of buildings (https://www.youtube.com/watch?v=SObzNdyRTBs), micromechanical devices (http://tinyurl.com/tinyprinter) and many things in between (https://www.youtube.com/watch?v=xVU4FLrsPXs).
16. Global map of community-based fab labs: https://www.fablabs.io/labs/map.
17. News article about Robert Williams being killed by an industrial robot: http://tinyurl.com/williamsaccident.
18. News article about Kenji Urada being killed by an industrial robot: http://tinyurl.com/uradaaccident.
19. News article about Volkswagen worker being killed by an industrial robot: http://tinyurl.com/baunatalaccident.
20. U.S. government report on worker fatalities: https://www.osha.gov/dep/fatcat/dep_fatcat.html.
21. Car accident fatality statistics: http://tinyurl.com/roadsafety2 and http://tinyurl.com/roadsafety3.
22. On the first Tesla autopilot fatality, see Andrew Buncombe, "Tesla Crash: Driver Who Died While on Autopilot Mode 'Was Watching Harry Potter,'"

Independent (July 1, 2016), http://tinyurl.com/teslacrashstory. For the report of the Office of Defects Investigation of the U.S. National Highway Traffic Safety Administration, see http://tinyurl.com/teslacrashreport.

23. For more about the *Herald of Free Enterprise* disaster, see R. B. Whittingham, *The Blame Machine: Why Human Error Causes Accidents* (Oxford, UK: Elsevier, 2004).
24. Documentary about the Air France 447 crash: https://www.youtube.com/watch?v=dpPkp8OGQFI; accident report: http://tinyurl.com/af447report; outside analysis: http://tinyurl.com/thomsonarticle.
25. Official report on the 2003 U.S.-Canadian blackout: http://tinyurl.com/uscanadablackout.
26. Final report of the President's Commission on the Accident at Three Mile Island: http://www.threemileisland.org/downloads/188.pdf.
27. Dutch study showing how AI can rival human radiologists at MRI-based diagnosis of prostate cancer: http://tinyurl.com/prostate-ai.
28. Stanford study showing how AI can best human pathologists at lung cancer diagnosis: http://tinyurl.com/lungcancer-ai.
29. Investigation of the Therac-25 radiation therapy accidents: http://tinyurl.com/theracfailure.
30. Report on lethal radiation overdoses in Panama caused by confusing user interface: http://tinyurl.com/cobalt60accident.
31. Study of adverse events in robotic surgery: https://arxiv.org/abs/1507.03518.
32. Article on number of deaths from bad hospital care: http://tinyurl.com/medaccidents.
33. Yahoo set a new standard for "big hack" when announcing a billion(!) of their user accounts had been breached: https://www.wired.com/2016/12/yahoo-hack-billion-users/.
34. *New York Times* article on acquittal and later conviction of KKK murderer: http://tinyurl.com/kkkacquittal.
35. The Danziger et al. 2011 study (http://www.pnas.org/content/108/17/6889.full), arguing that hungry judges are harsher, was criticized as flawed by Keren Weinshall-Margela and John Shapard (http://www.pnas.org/content/108/42/E833.full), but Danziger et al. insist that their claims remain valid (http://www.pnas.org/content/108/42/E834.full).
36. *Pro Publica* report on racial bias in recidivism-prediction software: http://tinyurl.com/robojudge.
37. Use of fMRI and other brain-scanning techniques as evidence in trials is highly controversial, as is the reliability of such techniques, although many teams claim accuracies better than 90%: http://journal.frontiersin.org/article/10.3389/fpsyg.2015.00709/full.
38. PBS made the movie *The Man Who Saved the World* about the incident where Vasili Arkhipov single-handedly prevented a Soviet nuclear strike: https://www.youtube.com/watch?v=4VPY2SgyG5w.
39. The story of how Stanislav Petrov dismissed warnings of a U.S. nuclear attack as a false alarm was turned into the movie *The Man Who Saved the World* (not to be confused with the movie by the same title in the previous note), and Petrov was honored at the United Nations and given the World Citizen Award: https://www.youtube.com/watch?v=IncSjwWQHMo.
40. Open letter from AI and robotics researchers about autonomous weapons: http://futureoflife.org/open-letter-autonomous-weapons/.

41. A U.S. official seemingly wanting a military AI arms race: http://tinyurl.com/workquote.
42. Study of wealth inequality in the United States since 1913: http://gabriel-zucman.eu/files/SaezZucman2015.pdf.
43. Oxfam report on global wealth inequality: http://tinyurl.com/oxfam2017.
44. For a great introduction to the hypothesis of technology-driven inequality, see Erik Brynjolfsson and Andrew McAfee, *The Second Machine Age: Work, Progress, and Prosperity in a Time of Brilliant Technologies* (New York: Norton, 2014).
45. Article in *The Atlantic* about falling wages for the less educated: http://tinyurl.com/wagedrop.
46. The data plotted are taken from Facundo Alvaredo, Anthony B. Atkinson, Thomas Piketty, Emmanuel Saez and Gabriel Zucman, *The World Wealth and Income Database* (http://www.wid.world), including capital gains.
47. Presentation by James Manyika showing income shifting from labor to capital: http://futureoflife.org/data/PDF/james_manyika.pdf.
48. Forecasts about future job automation from Oxford University (http://tinyurl.com/automationoxford) and McKinsey (http://tinyurl.com/automationmckinsey).
49. Video of robotic chef: https://www.youtube.com/watch?v=fE6i2OO6Y6s.
50. Marin Soljačić explored these options at the 2016 workshop Computers Gone Wild: Impact and Implications of Developments in Artificial Intelligence on Society: http://futureoflife.org/2016/05/06/computers-gone-wild/.
51. Andrew McAfee's suggestions for how to create more good jobs: http://futureoflife.org/data/PDF/andrew_mcafee.pdf.
52. In addition to many academic articles arguing that "this time is different" for technological unemployment, the video "Humans Need Not Apply" succinctly makes the same point: https://www.youtube.com/watch?v=7Pq-S557XQU.
53. U.S. Bureau of Labor Statistics: http://www.bls.gov/cps/cpsaat11.htm.
54. Argument that "this time is different" for technological unemployment: Federico Pistono, *Robots Will Steal Your Job, but That's OK* (2012), http://robotswillstealyourjob.com.
55. Changes in the U.S. horse population: http://tinyurl.com/horsedecline.
56. Meta-analysis showing how unemployment affects well-being: Maike Luhmann et al., "Subjective Well-Being and Adaptation to Life Events: A Meta-Analysis," *Journal of Personality and Social Psychology* 102, no. 3 (2012): 592; available online at https://www.ncbi.nlm.nih.gov/pmc/articles/PMC3289759.
57. Studies of what boosts people's sense of well-being: Angela Duckworth, Tracy Steen and Martin Seligman, "Positive Psychology in Clinical Practice," *Annual Review of Clinical Psychology* 1 (2005): 629–651, online at http://tinyurl.com/wellbeingduckworth. Weiting Ng and Ed Diener, "What Matters to the Rich and the Poor? Subjective Well-Being, Financial Satisfaction, and Postmaterialist Needs Across the World," *Journal of Personality and Social Psychology* 107, no. 2 (2014): 326, online at http://psycnet.apa.org/journals/psp/107/2/326. Kirsten Weir, "More than Job Satisfaction," *Monitor on Psychology* 44, no. 11 (December 2013), online at http://www.apa.org/monitor/2013/12/job-satisfaction.aspx.
58. Multiplying together about 10^{11} neurons, about 10^{4} connections per neuron and about one (10^{0}) firing per neuron each second might suggest that about 10^{15} FLOPS (1 petaFLOPS) suffice to simulate a human brain, but there are many poorly understood complications, including the detailed timing of firings

and the question of whether small parts of neurons and synapses need to be simulated too. IBM computer scientist Dharmendra Modha has estimated that 38 petaFLOPS are required (http://tinyurl.com/javln43), while neuroscientist Henry Markram has estimated that one needs about 1,000 petaFLOPS (http://tinyurl.com/6rpohqv). AI researchers Katja Grace and Paul Christiano have argued that the most costly aspect of brain simulation is not computation but *communication*, and that this too is a task in the ballpark of what the best current supercomputers can do: http://aiimpacts.org/about.

59. For an interesting estimate of the computational power of the human brain: Hans Moravec "When Will Computer Hardware Match the Human Brain?" *Journal of Evolution and Technology*, vol. 1 (1998).

Chapter 4

1. For a video of the first mechanical bird, see Markus Fischer, "A Robot That Flies like a Bird," TED Talk, July 2011, at https://www.ted.com/talks/a_robot_that_flies_like_a_bird.

Chapter 5

1. Ray Kurzweil, *The Singularity Is Near* (New York: Viking Press, 2005).
2. Ben Goertzel's "Nanny AI" scenario is described here: https://wiki.lesswrong.com/wiki/Nanny_AI.
3. For a discussion about the relationship between machines and humans, and whether machines are our slaves, see Benjamin Wallace-Wells, "Boyhood," *New York* magazine (May 20, 2015), online at http://tinyurl.com/aislaves.
4. Mind crime is discussed in Nick Bostrom's book *Superintelligence* and in more technical detail in this recent paper: Nick Bostrom, Allan Dafoe and Carrick Flynn, "Policy Desiderata in the Development of Machine Superintelligence" (2016), http://www.nickbostrom.com/papers/aipolicy.pdf.
5. Matthew Schofield, "Memories of Stasi Color Germans' View of U.S. Surveillance Programs,"*McClatchy DC Bureau* (June 26, 2013), online at http://www.mcclatchydc.com/news/nation-world/national/article24750439.html.
6. For thought-provoking reflections on how people can be incentivized to create outcomes that nobody wants, I recommend "Meditations on Moloch," http://slatestarcodex.com/2014/07/30/meditations-on-moloch.
7. For an interactive timeline of close calls when nuclear war might have started by accident, see Future of Life Institute, "Accidental Nuclear War: A Timeline of Close Calls," online at http://tinyurl.com/nukeoops.
8. For compensation payments made to U.S. nuclear testing victims, see U.S. Department of Justice website, "Awards to Date 4/24/2015," at https://www.justice.gov/civil/awards-date-04242015.
9. *Report of the Commission to Assess the Threat to the United States from Electromagnetic Pulse (EMP) Attack*, April 2008, available online at http://www.empcommission.org/docs/A2473-EMP_Commission-7MB.pdf.
10. Independent research by both U.S. and Soviet scientists alerted Reagan and Gorbachev to the risk of nuclear winter: P. J. Crutzen and J. W. Birks, "The

Atmosphere After a Nuclear War: Twilight at Noon," *Ambio* 11, no. 2/3 (1982): 114–125. R. P. Turco, O. B. Toon, T. P. Ackerman, J. B. Pollack and C. Sagan, "Nuclear Winter: Global Consequences of Multiple Nuclear Explosions," *Science* 222 (1983): 1283–1292. V. V. Aleksandrov and G. L. Stenchikov, "On the Modeling of the Climatic Consequences of the Nuclear War," *Proceeding on Applied Mathematics* (Moscow: Computing Centre of the USSR Academy of Sciences, 1983), 21. A. Robock, "Snow and Ice Feedbacks Prolong Effects of Nuclear Winter," *Nature* 310 (1984): 667–670.

11. Calculation of climate effects of global nuclear war: A. Robock, L. Oman and L. Stenchikov, "Nuclear Winter Revisited with a Modern Climate Model and Current Nuclear Arsenals: Still Catastrophic Consequences," *Journal of Geophysical Research* 12 (2007): D13107.

Chapter 6

1. For more information, see Anders Sandberg, "Dyson Sphere FAQ," at http://www.aleph.se/nada/dysonFAQ.html.
2. Freeman Dyson's seminal paper on his eponymous spheres: Freeman Dyson, "Search for Artificial Stellar Sources of Infrared Radiation," *Science*, vol. 131 (1959): 1667–1668.
3. Louis Crane and Shawn Westmoreland explain their proposed black hole engine in "Are Black Hole Starships Possible?," at http://arxiv.org/pdf/0908.1803.pdf.
4. For a nice infographic from CERN summarizing known elementary particles, see http://tinyurl.com/cernparticles.
5. This unique video of a non-nuclear Orion prototype illustrates the idea of nuclear-bomb-powered rocket propulsion: https://www.youtube.com/watch?v=E3Lxx2VAYi8.
6. Here's a pedagogical introduction to laser sailing: Robert L. Forward, "Roundtrip Interstellar Travel Using Laser-Pushed Lightsails," *Journal of Spacecraft and Rockets* 21, no. 2 (March–April 1984), available online at http://www.lunarsail.com/LightSail/rit-1.pdf.
7. Jay Olson analyzes cosmically expanding civilizations in "Homogeneous Cosmology with Aggressively Expanding Civilizations," *Classical and Quantum Gravity* 32 (2015), available online at http://arxiv.org/abs/1411.4359.
8. The first thorough scientific analysis of our far future: Freeman J. Dyson, "Time Without End: Physics and Biology in an Open Universe," *Reviews of Modern Physics* 51, no. 3 (1979): 447, available online at http://blog.regehr.org/extra_files/dyson.pdf.
9. Seth Lloyd's above-mentioned formula told us that performing a computational operation during a time interval τ costs an energy $E \geq h/4\tau$, where h is Planck's constant. If we want to get N operations done one after the other (in series) during a time T, then $\tau = T/N$, so $E/N \geq hN/4T$, which tells us that we can perform $N \leq 2\sqrt{ET/h}$ serial operations using energy E and time T. So both energy and time are resources that it helps having lots of. If you split your energy between n different parallel computations, they can run more slowly and efficiently, giving $N \leq 2\sqrt{ETn/h}$. Nick Bostrom estimates that simulating a 100-year human life requires about $N = 10^{27}$ operations.

10. If you want to see a careful argument for why the origin of life may require a very rare fluke, placing our nearest neighbors over 10^{1000} meters away, I recommend this video by Princeton physicist and astrobiologist Edwin Turner: "Improbable Life: An Unappealing but Plausible Scenario for Life's Origin on Earth," at https://www.youtube.com/watch?v=Bt6n6Tu1beg.
11. Essay by Martin Rees on the search for extraterrestrial intelligence: https://www.edge.org/annual-question/2016/response/26665.

Chapter 7

1. A popular discussion of Jeremy England's work on "dissipation-driven adaptation" can be found in Natalie Wolchover, "A New Physics Theory of Life," *Scientific American* (January 28, 2014), available online at https://www.scientificamerican.com/article/a-new-physics-theory-of-life/. Ilya Prigogine and Isabelle Stengers's *Order Out of Chaos: Man's New Dialogue with Nature* (New York: Bantam, 1984) lays many of the foundations for this.
2. For more on feelings and their physiological roots: William James, *Principles of Psychology* (New York: Henry Holt & Co., 1890); Robert Ornstein, *Evolution of Consciousness: The Origins of the Way We Think* (New York: Simon & Schuster, 1992); António Damásio, *Descartes' Error: Emotion, Reason, and the Human Brain* (New York: Penguin, 2005); and António Damásio, *Self Comes to Mind: Constructing the Conscious Brain* (New York: Vintage, 2012).
3. Eliezer Yudkowsky has discussed aligning the goals of friendly AI not with our present goals, but with our *coherent extrapolated volition* (CEV). Loosely speaking this is defined as what an idealized version of us would want if we knew more, thought faster and were more the people we wished we were. Yudkowsky began criticizing CEV shortly after publishing it in 2004 (http://intelligence.org/files/CEV.pdf), both for being hard to implement and because it's unclear whether it would converge to anything well-defined.
4. In the inverse reinforcement-learning approach, a core idea is that the AI is trying to maximize not its own goal-satisfaction, but that of its human owner. It therefore has incentive to be cautious when it's unclear about what its owner wants, and to do its best to find out. It should also be fine with its owner switching it off, since that would imply that it had misunderstood what its owner really wanted.
5. Steve Omohundro's paper on AI goal emergence, "The Basic AI Drives," can be found online at http://tinyurl.com/omohundro2008. Originally published in *Artificial General Intelligence 2008: Proceedings of the First AGI Conference*, ed. Pei Wang, Ben Goertzel and Stan Franklin (Amsterdam: IOS, 2008), 483–492.
6. A thought-provoking and controversial book on what happens when intelligence is used to blindly obey orders without questioning their ethical basis: Hannah Arendt, *Eichmann in Jerusalem: A Report on the Banality of Evil* (New York: Penguin, 1963). A related dilemma applies to a recent proposal by Eric Drexler (http://www.fhi.ox.ac.uk/reports/2015-3.pdf) to keep superintelligence under control by compartmentalizing it into simple pieces, none of which understand the whole picture. If this works, this could again provide an incredibly powerful tool without an intrinsic moral compass, implementing its owner's every whim

without any moral qualms. This would be reminiscent of a compartmentalized bureaucracy in a dystopian dictatorship: one part builds weapons without knowing how they'll be used, another executes prisoners without knowing why they were convicted, and so on.

7. A modern variant of the Golden Rule is John Rawls' idea that a hypothetical situation is fair if nobody would change it without knowing in advance which person in it they'd be.
8. For example, the IQs of many of Hitler's top officials were found to be quite high. See "How Accurate Were the IQ Scores of the High-Ranking Third Reich Officials Tried at Nuremberg?," *Quora*, available online at http://tinyurl.com/nurembergiq.

Chapter 8

1. The entry on consciousness by Stuart Sutherland is quite amusing: *Macmillan Dictionary of Psychology* (London: Macmillan, 1989).
2. Erwin Schrödinger, one of the founding fathers of quantum mechanics, made this remark in his book *Mind and Matter* while contemplating the *past*—and what would have happened if conscious life never evolved in the first place. On the other hand, the rise of AI raises the logical possibility that we may end up with a play for empty benches in the *future*.
3. The *Stanford Encyclopedia of Philosophy* gives an extensive survey of different definitions and uses of the word "consciousness": http://tinyurl.com/stanfordconsciousness.
4. Yuval Noah Harari, *Homo Deus: A Brief History of Tomorrow* (New York: HarperCollins, 2017): 116.
5. This is an excellent introduction to System 1 and System 2 from a pioneer: Daniel Kahneman, *Thinking, Fast and Slow* (New York: Farrar, Straus & Giroux, 2011).
6. See Christof Koch, *The Quest for Consciousness: A Neurobiological Approach* (New York: W. H. Freeman, 2004).
7. Perhaps we're only aware of a tiny fraction (say 10–50 bits) of the information that enters our brain each second: K. Küpfmüller, 1962, "Nachrichtenverarbeitung im Menschen," in *Taschenbuch der Nachrichtenverarbeitung*, ed. K. Steinbuch (Berlin: Springer-Verlag, 1962): 1481–1502. T. Nørretranders, *The User Illusion: Cutting Consciousness Down to Size* (New York: Viking, 1991).
8. Michio Kaku, *The Future of the Mind: The Scientific Quest to Understand, Enhance, and Empower the Mind* (New York: Doubleday, 2014); Jeff Hawkins and Sandra Blakeslee, *On Intelligence* (New York: Times Books, 2007); Stanislas Dehaene, Michel Kerszberg and Jean-Pierre Changeux, "A Neuronal Model of a Global Workspace in Effortful Cognitive Tasks," *Proceedings of the National Academy of Sciences* 95 (1998): 14529–14534.
9. Video celebrating Penfield's famous "I can smell burnt toast" experiment: https://www.youtube.com/watch?v=mSN86kphL68. Sensorimotor cortex details: Elaine Marieb and Katja Hoehn, *Anatomy & Physiology*, 3rd ed. (Upper Saddle River, NJ: Pearson, 2008), 391–395.
10. The study of neural correlates of consciousness (NCCs) has become quite mainstream in the neuroscience community in recent years—see, e.g., Geraint Rees,

Gabriel Kreiman, and Christof Koch, "Neural Correlates of Consciousness in Humans," *Nature Reviews Neuroscience* 3 (2002): 261–270, and Thomas Metzinger, *Neural Correlates of Consciousness: Empirical and Conceptual Questions* (Cambridge, MA: MIT Press, 2000).

11. How continuous flash suppression works: Christof Koch, *The Quest for Consciousness: A Neurobiological Approach* (New York: W. H. Freeman, 2004); Christof Koch and Naotsugu Tsuchiya, "Continuous Flash Suppression Reduces Negative Afterimages," *Nature Neuroscience* 8 (2005): 1096–1101.
12. Christof Koch, Marcello Massimini, Melanie Boly and Giulio Tononi, "Neural Correlates of Consciousness: Progress and Problems," *Nature Reviews Neuroscience* 17 (2016): 307.
13. See Koch, *The Quest for Consciousness*, p. 260, and further discussion in the *Stanford Encyclopedia of Philosophy*, http://tinyurl.com/consciousnessdelay.
14. On synchronization of conscious perception: David Eagleman, *The Brain: The Story of You* (New York: Pantheon, 2015), and *Stanford Encyclopedia of Philosophy*, http://tinyurl.com/consciousnesssync.
15. Benjamin Libet, *Mind Time: The Temporal Factor in Consciousness* (Cambridge, MA: Harvard University Press, 2004); Chun Siong Soon, Marcel Brass, Hans-Jochen Heinze and John-Dylan Haynes, "Unconscious Determinants of Free Decisions in the Human Brain," *Nature Neuroscience* 11 (2008): 543–545, online at http://www.nature.com/neuro/journal/v11/n5/full/nn.2112.html.
16. Examples of recent theoretical approaches to consciousness:
    - Daniel Dennett, *Consciousness Explained* (Back Bay Books, 1992)
    - Bernard Baars, *In the Theater of Consciousness: The Workspace of the Mind* (New York: Oxford University Press, 2001)
    - Christof Koch, *The Quest for Consciousness: A Neurobiological Approach* (New York: W. H. Freeman, 2004)
    - Gerald Edelman and Giulio Tononi, *A Universe of Consciousness: How Matter Becomes Imagination* (New York: Hachette, 2008)
    - António Damásio, *Self Comes to Mind: Constructing the Conscious Brain* (New York: Vintage, 2012)
    - Stanislas Dehaene, *Consciousness and the Brain: Deciphering How the Brain Codes Our Thoughts* (New York: Viking, 2014)
    - Stanislas Dehaene, Michel Kerszberg and Jean-Pierre Changeux, "A Neuronal Model of a Global Workspace in Effortful Cognitive Tasks," *Proceedings of the National Academy of Sciences* 95 (1998): 14529–14534
    - Stanislas Dehaene, Lucie Charles, Jean-Rémi King and Sébastien Marti, "Toward a Computational Theory of Conscious Processing," *Current Opinion in Neurobiology* 25 (2014): 760–784
17. Thorough discussion of different uses of the term "emergence" in physics and philosophy by David Chalmers: http://cse3521.artifice.cc/Chalmers-Emergence.pdf.
18. Me arguing that consciousness is the way information feels when being processed in certain complex ways: https://arxiv.org/abs/physics/0510188, https://arxiv.org/abs/0704.0646, Max Tegmark, *Our Mathematical Universe* (New York: Knopf, 2014). David Chalmers expresses a related sentiment in his 1996 book *The Conscious Mind:* "Experience is information from the inside; physics is information from the outside."
19. Adenauer Casali et al., "A Theoretically Based Index of Consciousness Inde-

pendent of Sensory Processing and Behavior," *Science Translational Medicine* 5 (2013): 198ra105, online at http://tinyurl.com/zapzip.

20. Integrated information theory doesn't work for continuous systems:
    - https://arxiv.org/abs/1401.1219
    - http://journal.frontiersin.org/article/10.3389/fpsyg.2014.00063/full
    - https://arxiv.org/abs/1601.02626
21. Interview with Clive Wearing, whose short-term memory is only about 30 seconds: https://www.youtube.com/watch?v=WmzU47i2xgw.
22. Scott Aaronson IIT critique: http://www.scottaaronson.com/blog/?p=1799.
23. Cerrullo IIT critique, arguing that integration isn't a sufficient condition for consciousness: http://tinyurl.com/cerrullocritique.
24. IIT prediction that simulated humans will be zombies: http://rstb.royalsocietypublishing.org/content/370/1668/20140167.
25. Shanahan critique of IIT: https://arxiv.org/pdf/1504.05696.pdf.
26. Blindsight: http://tinyurl.com/blindsight-paper.
27. Perhaps we're only aware of a tiny fraction (say 10–50 bits) of the information that enters our brain each second: Küpfmüller, "Nachrichtenverarbeitung im Menschen"; Nørretranders, *The User Illusion*.
28. The case for and against "consciousness without access": Victor Lamme, "How Neuroscience Will Change Our View on Consciousness," *Cognitive Neuroscience* (2010): 204–220, online at http://www.tandfonline.com/doi/abs/10.1080/17588921003731586.
29. "Selective Attention Test," at https://www.youtube.com/watch?v=vJG698U2Mvo.
30. See Lamme, "How Neuroscience Will Change Our View on Consciousness," n. 28.
31. This and other related issues are discussed in detail in Daniel Dennett's book *Consciousness Explained*.
32. See Kahneman, *Thinking, Fast and Slow*, cited in n. 5.
33. The *Stanford Encyclopedia of Philosophy* reviews the free will controversy: https://plato.stanford.edu/entries/freewill.
34. Video of Seth Lloyd explaining why an AI will feel like it has free will: https://www.youtube.com/watch?v=Epj3DF8jDWk.
35. See Steven Weinberg, *Dreams of a Final Theory: The Search for the Fundamental Laws of Nature* (New York: Pantheon, 1992).
36. The first thorough scientific analysis of our far future: Freeman J. Dyson, "Time Without End: Physics and Biology in an Open Universe," *Reviews of Modern Physics* 51, no. 3 (1979): 447, available online at http://blog.regehr.org/extra_files/dyson.pdf.

Epilogue

1. The open letter (http://futureoflife.org/ai-open-letter) that emerged from the Puerto Rico conference argued that research on how to make AI systems robust and beneficial is both important and timely, and that there are concrete research directions that can be pursued today, as exemplified in this research-priorities document: http://futureoflife.org/data/documents/research_priorities.pdf.

2. My video interview with Elon Musk about AI safety can be found on YouTube at https://www.youtube.com/watch?v=rBw0eoZTY-g.
3. Here's a nice video compilation of almost all SpaceX rocket landing attempts, culminating with the first successful ocean landing: https://www.youtube.com/watch?v=AllaFzIPaG4.
4. Elon Musk tweets about our AI-safety grant competition: https://twitter.com/elonmusk/status/555743387056226304.
5. Elon Musk tweets about our AI-safety-endorsing open letter: https://twitter.com/elonmusk/status/554320532133650432.
6. Erik Sofge in "An Open Letter to Everyone Tricked into Fearing Artificial Intelligence" (*Popular Science*, January 14, 2015) pokes fun at the scaremongering news coverage of our open letter: http://www.popsci.com/open-letter-everyone-tricked-fearing-ai.
7. Elon Musk tweets about his big donation to the Future of Life Institute and the world of AI-safety researchers: https://twitter.com/elonmusk/status/555743387056226304.
8. For more about the Partnership on AI to benefit people and society, see their website: https://www.partnershiponai.org.
9. Some examples of recent reports expressing opinions about AI: One Hundred Year Study on Artificial Intelligence, Report of the 2015 Study Panel, "Artificial Intelligence and Life in 2030" (September 2016), at http://tinyurl.com/stanfordai; White House report on the future of AI: http://tinyurl.com/obamaAIreport; White House report on AI and jobs: http://tinyurl.com/AIjobsreport; IEEE report on AI and human well-being, "Ethically Aligned Design, Version 1" (December 13, 2016), at http://standards.ieee.org/develop/indconn/ec/ead_v1.pdf; road map for U.S. Robotics: http://tinyurl.com/roboticsmap.
10. Among the principles that didn't make the final cut, one of my favorites was this one: "Consciousness caution: There being no consensus, we should avoid strong assumptions as to whether or not advanced AI would possess or require consciousness or feelings." It went through many iterations, and in the last one, the controversial word "consciousness" was replaced by "subjective experience"—but this principle nonetheless got only 88% approval, just barely falling short of the 90% cutoff.
11. Discussion panel on superintelligence with Elon Musk and other great minds: http://tinyurl.com/asilomarAI.

Index

ALLEN LANE
an imprint of
PENGUIN BOOKS

Also Published

Michael Pollan, *How to Change Your Mind: The New Science of Psychedelics*

David Christian, *Origin Story: A Big History of Everything*

Judea Pearl and Dana Mackenzie, *The Book of Why: The New Science of Cause and Effect*

David Graeber, *Bullshit Jobs: A Theory*

Serhii Plokhy, *Chernobyl: History of a Tragedy*

Michael McFaul, *From Cold War to Hot Peace: The Inside Story of Russia and America*

Paul Broks, *The Darker the Night, the Brighter the Stars: A Neuropsychologist's Odyssey*

Lawrence Wright, *God Save Texas: A Journey into the Future of America*

John Gray, *Seven Types of Atheism*

Carlo Rovelli, *The Order of Time*

Mariana Mazzucato, *The Value of Everything: Making and Taking in the Global Economy*

Richard Vinen, *The Long '68: Radical Protest and Its Enemies*

Kishore Mahbubani, *Has the West Lost It?: A Provocation*

John Lewis Gaddis, *On Grand Strategy*

Richard Overy, *The Birth of the RAF, 1918: The World's First Air Force*

Francis Pryor, *Paths to the Past: Encounters with Britain's Hidden Landscapes*

Helen Castor, *Elizabeth I: A Study in Insecurity*

Ken Robinson and Lou Aronica, *You, Your Child and School*

Leonard Mlodinow, *Elastic: Flexible Thinking in a Constantly Changing World*

Nick Chater, *The Mind is Flat: The Illusion of Mental Depth and The Improvised Mind*

Michio Kaku, *The Future of Humanity: Terraforming Mars, Interstellar Travel, Immortality, and Our Destiny Beyond*

Thomas Asbridge, *Richard I: The Crusader King*

Richard Sennett, *Building and Dwelling: Ethics for the City*

Nassim Nicholas Taleb, *Skin in the Game: Hidden Asymmetries in Daily Life*

Steven Pinker, *Enlightenment Now: The Case for Reason, Science, Humanism and Progress*

Steve Coll, *Directorate S: The C.I.A. and America's Secret Wars in Afghanistan, 2001 - 2006*

Jordan B. Peterson, *12 Rules for Life: An Antidote to Chaos*

Bruno Maçães, *The Dawn of Eurasia: On the Trail of the New World Order*

Brock Bastian, *The Other Side of Happiness: Embracing a More Fearless Approach to Living*

Ryan Lavelle, *Cnut: The North Sea King*

Tim Blanning, *George I: The Lucky King*

Thomas Cogswell, *James I: The Phoenix King*

Pete Souza, *Obama, An Intimate Portrait: The Historic Presidency in Photographs*

Robert Dallek, *Franklin D. Roosevelt: A Political Life*

Norman Davies, *Beneath Another Sky: A Global Journey into History*

Ian Black, *Enemies and Neighbours: Arabs and Jews in Palestine and Israel, 1917-2017*

Martin Goodman, *A History of Judaism*

Shami Chakrabarti, *Of Women: In the 21st Century*

Stephen Kotkin, *Stalin, Vol. II: Waiting for Hitler, 1928-1941*

Lindsey Fitzharris, *The Butchering Art: Joseph Lister's Quest to Transform the Grisly World of Victorian Medicine*

Serhii Plokhy, *Lost Kingdom: A History of Russian Nationalism from Ivan the Great to Vladimir Putin*

Mark Mazower, *What You Did Not Tell: A Russian Past and the Journey Home*

Lawrence Freedman, *The Future of War: A History*

Niall Ferguson, *The Square and the Tower: Networks, Hierarchies and the Struggle for Global Power*

Matthew Walker, *Why We Sleep: The New Science of Sleep and Dreams*

Edward O. Wilson, *The Origins of Creativity*

John Bradshaw, *The Animals Among Us: The New Science of Anthropology*

David Cannadine, *Victorious Century: The United Kingdom, 1800-1906*

Leonard Susskind and Art Friedman, *Special Relativity and Classical Field Theory*

Maria Alyokhina, *Riot Days*

Oona A. Hathaway and Scott J. Shapiro, *The Internationalists: And Their Plan to Outlaw War*

Chris Renwick, *Bread for All: The Origins of the Welfare State*

Anne Applebaum, *Red Famine: Stalin's War on Ukraine*

Richard McGregor, *Asia's Reckoning: The Struggle for Global Dominance*

Chris Kraus, *After Kathy Acker: A Biography*

Clair Wills, *Lovers and Strangers: An Immigrant History of Post-War Britain*

Odd Arne Westad, *The Cold War: A World History*

Max Tegmark, *Life 3.0: Being Human in the Age of Artificial Intelligence*

Jonathan Losos, *Improbable Destinies: How Predictable is Evolution?*

Chris D. Thomas, *Inheritors of the Earth: How Nature Is Thriving in an Age of Extinction*

Chris Patten, *First Confession: A Sort of Memoir*

James Delbourgo, *Collecting the World: The Life and Curiosity of Hans Sloane*

Naomi Klein, *No Is Not Enough: Defeating the New Shock Politics*

Ulrich Raulff, *Farewell to the Horse: The Final Century of Our Relationship*

Slavoj Žižek, *The Courage of Hopelessness: Chronicles of a Year of Acting Dangerously*

Patricia Lockwood, *Priestdaddy: A Memoir*

Ian Johnson, *The Souls of China: The Return of Religion After Mao*

Stephen Alford, *London's Triumph: Merchant Adventurers and the Tudor City*

Hugo Mercier and Dan Sperber, *The Enigma of Reason: A New Theory of Human Understanding*

Stuart Hall, *Familiar Stranger: A Life Between Two Islands*

Allen Ginsberg, *The Best Minds of My Generation: A Literary History of the Beats*

Sayeeda Warsi, *The Enemy Within: A Tale of Muslim Britain*

Alexander Betts and Paul Collier, *Refuge: Transforming a Broken Refugee System*

Robert Bickers, *Out of China: How the Chinese Ended the Era of Western Domination*

Erica Benner, *Be Like the Fox: Machiavelli's Lifelong Quest for Freedom*

William D. Cohan, *Why Wall Street Matters*

David Horspool, *Oliver Cromwell: The Protector*

Daniel C. Dennett, *From Bacteria to Bach and Back: The Evolution of Minds*

Derek Thompson, *Hit Makers: How Things Become Popular*

Harriet Harman, *A Woman's Work*

Wendell Berry, *The World-Ending Fire: The Essential Wendell Berry*

Daniel Levin, *Nothing but a Circus: Misadventures among the Powerful*

Stephen Church, *Henry III: A Simple and God-Fearing King*

Pankaj Mishra, *Age of Anger: A History of the Present*

Graeme Wood, *The Way of the Strangers: Encounters with the Islamic State*

Michael Lewis, *The Undoing Project: A Friendship that Changed the World*

John Romer, *A History of Ancient Egypt, Volume 2: From the Great Pyramid to the Fall of the Middle Kingdom*

Andy King, *Edward I: A New King Arthur?*

Thomas L. Friedman, *Thank You for Being Late: An Optimist's Guide to Thriving in the Age of Accelerations*

John Edwards, *Mary I: The Daughter of Time*

Grayson Perry, *The Descent of Man*

Deyan Sudjic, *The Language of Cities*

Norman Ohler, *Blitzed: Drugs in Nazi Germany*

Carlo Rovelli, *Reality Is Not What It Seems: The Journey to Quantum Gravity*

Catherine Merridale, *Lenin on the Train*

Susan Greenfield, *A Day in the Life of the Brain: The Neuroscience of Consciousness from Dawn Till Dusk*

Christopher Given-Wilson, *Edward II: The Terrors of Kingship*

Emma Jane Kirby, *The Optician of Lampedusa*

Minoo Dinshaw, *Outlandish Knight: The Byzantine Life of Steven Runciman*

Candice Millard, *Hero of the Empire: The Making of Winston Churchill*